Beginning iOS 5 Games Development

Using the iOS 5 SDK for iPad, iPhone, and iPod Touch

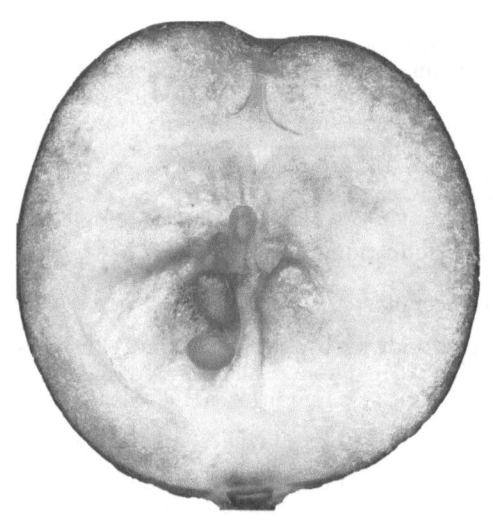

Lucas Jordan

Apress®

Beginning iOS 5 Games Development: Using the iOS 5 SDK for iPad, iPhone, and iPod Touch

ISBN 978-1-4302-3710-5

ISBN 978-1-4302-3711-2 (eBook)

President and Publisher: Paul Manning
Lead Editor: James Markham
Technical Reviewer: Tony Hillerson
Editorial Board: Steve Anglin, Mark Beckner, Ewan Buckingham, Gary Cornell,
    Morgan Engel, Jonathan Gennick, Jonathan Hassell, Robert Hutchinson, Michelle
    Lowman, James Markham, Matthew Moodie, Jeff Olson, Jeffrey Pepper, Douglas Pundick,
    Ben Renow-Clarke, Dominic Shakeshaft, Gwenan Spearing, Matt Wade, Tom Welsh
Coordinating Editor: Corbin Collins
Copy Editor: Tracy Brown
Compositor: MacPS, LLC
Indexer: BIM Indexing & Proofreading Services
Artist: SPi Global
Cover Designer: Anna Ishchenko

Distributed to the book trade worldwide by Springer Science+Business Media, LLC., 233 Spring Street, 6th Floor, New York, NY 10013. Phone 1-800-SPRINGER, fax (201) 348-4505, e-mail orders-ny@springer-sbm.com, or visit www.springeronline.com.

For information on translations, please e-mail rights@apress.com, or visit www.apress.com.

Apress and friends of ED books may be purchased in bulk for academic, corporate, or promotional use. eBook versions and licenses are also available for most titles. For more information, reference our Special Bulk Sales–eBook Licensing web page at www.apress.com/bulk-sales.

Any source code or other supplementary materials referenced by the author in this text is available to readers at www.apress.com. For detailed information about how to locate your book's source code, go to http://www.apress.com/source-code/.

To those who make great games.

Contents at a Glance

Contents

About the Author

 **Lucas L. Jordan** is a lifelong computer enthusiast who has worked for many years as a developer focusing on user interface design. He is the author of *JavaFX Special Effects: Taking Java RIA to the Extreme with Animation, Multimedia, and Game Elements* and the co-author of *Practical Android Projects*, both by Apress. Lucas is interested in mobile application development in its many forms. When the time is right, he will commit himself fulltime to game development at ClayWare Games, LLC.

About the Technical Reviewer

 **Tony Hillerson** is a software architect at EffectiveUI. He graduated from Ambassador University with a BA in Management Information Systems. On any given day, Tony might be working with Android, Rails, Objective-C, Java, Flex, or shell scripts. He has been interested in developing for Android since the early betas. Hillerson has created Android screencasts, has spoken about Android at conferences, and has served as technical reviewer on Android books. He also sometimes gets to write Android code.

He is interested in all levels of usability and experience design, from the database to the server to the glass.

In his free time, Hillerson enjoys playing the bass, playing World of Warcraft, and making electronic music. Tony lives outside Denver, Colorado, with his wife and two sons.

Acknowledgments

Without Corbin Collins this book would have never been completed. My wife and family deserve praise for helping me find the time to get this work done. Thank you, Tony Hillerson, for providing the technical editing. Lastly, special thanks to the rest of the Apress staff that helped get this book out the door in time for the iOS 5 release.

Introduction

This book is an introduction to game development for Apple's iOS devices. I hope you are reading this because you are excited about developing a cool new game for the iPhone or iPad. It is the goal of this book to educate the reader on the many facets of building a game. The topics range from the mundane tasks of organizing your start screen to the subtle math required to get an animation just right.

By the end of this book, you will be familiar with not just the details of animating images, but also the many other details that go into a game that should be considered before development even starts. For example, if you know you want to include Game Center in your application, it makes sense to understand Game Center and how that affects your design.

The book walks you through many examples, each building on the other to create a complete picture of what an iOS game looks like. To facilitate this, I have written a complete (though simple) game that will serve as context in the examples, as well as give you a complete game to use as a roadmap in your own game development. The game is called Belt Commander and is shown in Figure 1.

In Figure 1, we see the game Belt Commander's start screen and a shot of the game in action. This game includes elements from every chapter in this book, and I hope it is a useful tool.

This game does not have the polish required for me to put it in the App Store, but it will serve as a starting point for a version of the game I do intend to publish. I just want to point this out, because I hope it highlights the practical approach of this book. The lessons I learned building Belt Commander have been captured in these pages. You can find the published game at claywaregames.com.

If you have questions about the code in this book, or about iOS development in general, please feel free to contact me directly at lucasjordan@gmail.com.

Figure 1. Belt Commander

Chapter-by-Chapter Overview

Each chapter in this book covers a particular topic pertinent to game development. The source code in each chapter is taken from a working sample project. Some chapters have their own projects, while others share a single, bigger project.

Chapter 1

Chapter 1 provides a walkthrough of setting up a simple project using Xcode. Using the code provided with this book, you will make your first simple game.

Chapter 2

Chapter 2 continues our exploration of Xcode and the game project. In this chapter, we talk about `UIViewControllers` and how they are designed to encapsulate functionality into reusable components. We will also address the issue of the different devices and orientations and applications you must work with, as shown in Figure 2.

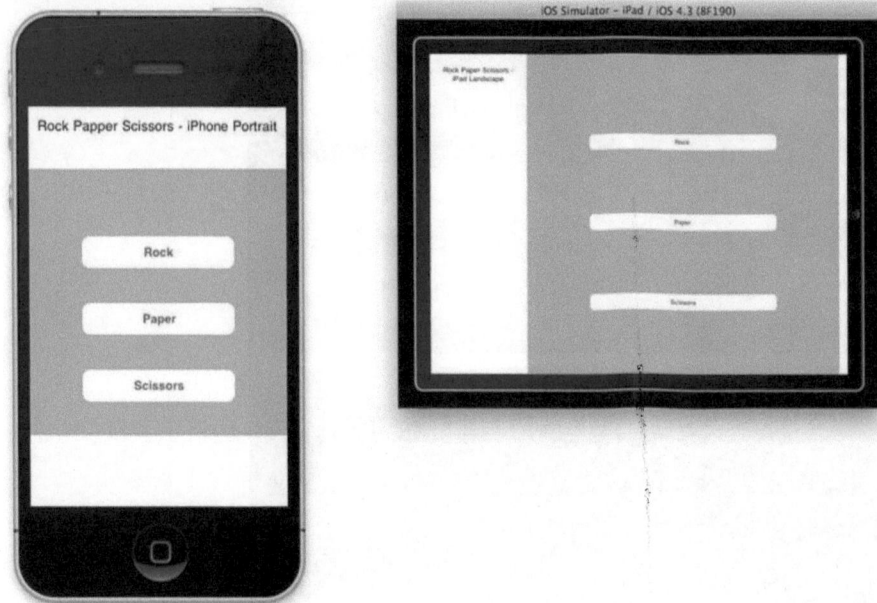

Figure 2. A simple game in portrait view on the iPhone and landscape view on the iPad

In Figure 2, we see a simple Rock, Paper, Scissors game being run on both the iPad and the iPhone. On the iPhone, the application is running in portrait view, while on the iPad the game is running in landscape view. Not every game needs to support all devices and orientations, but some do. In this chapter, you will learn how to get a project off on the right foot, so you don't have to fight with it down the road.

Chapter 3

Chapter 3 is concerned with understanding the life cycle of an iOS application. It will cover how an application is initialized, how to navigate between a number of different views, and how to handle the application going into the background. Figure 3 shows the different views in the project we will be working with.

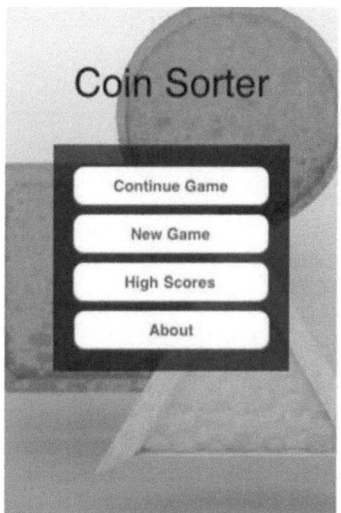

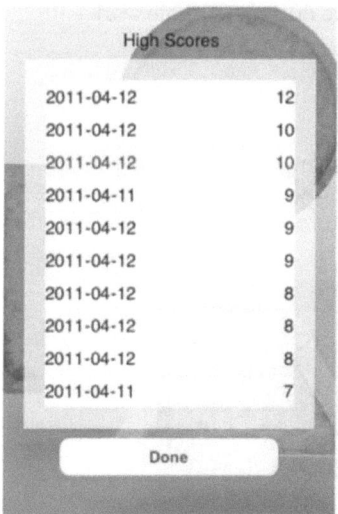
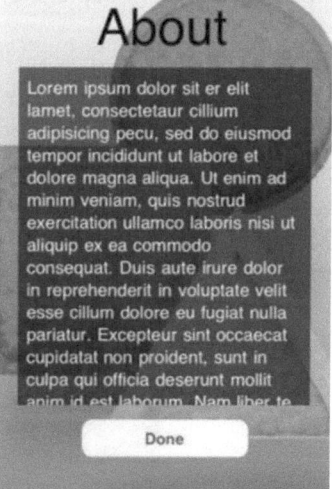

Figure 3. *The four views of the game Coin Sorter*

In Figure 3, we see the four views of the game Coin Sorter. You will learn in this chapter how to design the navigation between these views, how to persist game state and high scores, and you will look at the details of preserving state when an applications enter the background. This will give you the context in which you can create games.

Chapter 4

In Chapter 4, we dive into the details of creating the fun part of the game. Up until this point, we have been mostly concerned with the stuff that surrounds the game itself. In this chapter, we will look at the implementation of the Cion Sorter game. You will learn how to build a game that is driven by user input. Figure 4 shows a preview of the type of things you will be learning.

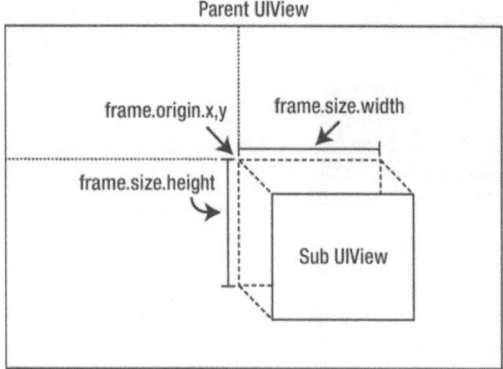

Figure 4. *A preview of how UIViews work*

In Figure 4, we see a visual description of the location of a view is described in iOS. Up until Chapter 4, we have glossed over the exact details of how views are positioned in the screen. In Chapter 4, we take our first in-depth look at the class UIView and how to use Core Graphics to describe the location of UIView. The techniques you learn will serve you in building the input-driven game and come into play in the following chapters, when we do considerable work with UIView and UIImageView.

Chapter 5

The game built in Chapter 4 uses an architecture suitable for "casual" games. Chapter 5 introduces the idea of a frame-by-frame game, where our code is responsible for drawing every frame of an animation. Figure 5 shows an overview of the architecture we will be creating and the sample animation that results.

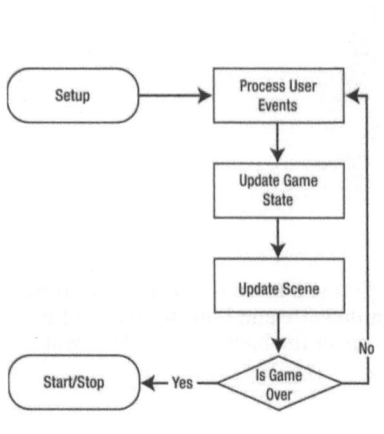

Figure 5. *Architecture of frame-by-frame game with sample game*

In Figure 5, we see on the left a flow chart describing how a frame-by-frame game works. We will work through each step to illustrate how to create a sample game like the one you see on the right. You will also be introduced to the space theme that will dominate the rest of the book, because in this chapter you learn the core principles for building our example game, Belt Commander.

Once we have a basic idea of how a frame-by-frame game works, we move on to the next chapter, where we start creating classes that provide the core functionality required to make a game.

Chapter 6

In Chapter 6, we take the core principles from Chapter 5 and wrap up that functionality into some reusable classes. In effect, we start building a basic game engine. We create a master class for managing the game and introduce the concepts of actors and behaviors. Figure 6 shows an example of an actor.

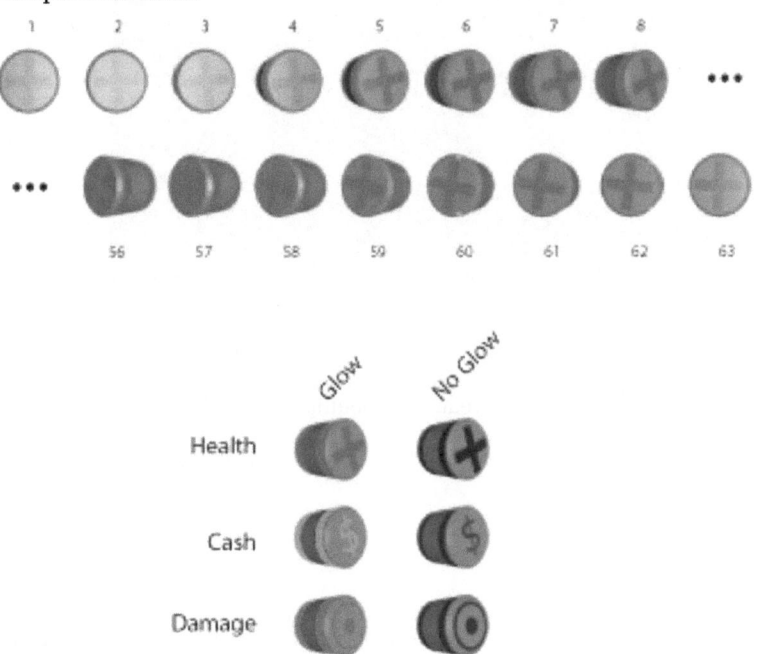

Figure 6. A Sample actor: The power-up

In Figure 6, we see the images that make up the actor power-up. An actor is an object in a game that encapsulates its behavior and visual representation. By the end of this chapter, you will have a simple pattern for creating a game, adding actors to it, and watching the actors animate onscreen in accordance with their behavior. In this chapter, we identify the core elements of building this style of game.

Chapter 7

Chapter 7 takes the framework from Chapter 6 and adds a few additional features such as actors drawn with vectors and particles. In the process of learning about how to create these new types of actors, we build a couple of compelling examples, one of which is shown in Figure 7.

Figure 7. An example of actors composed of a particle system

In Figure 7, we see three comets flying across the screen. Each comet is composed of many small actors called particles. Particles are used to create compelling visual effects that would be hard to capture with a single image or animation.

By the end of this chapter, we will have several examples of setting up a game and creating custom actors to populate it with.

Chapter 8

Having established how to create a scene populated with custom actors in the previous two chapters, we are ready to look at user interactions. Up until this point, the examples have pretty much just run on their own, or with very limited user actions. In this chapter, we will explore all of the user input options, as shown in Figure 8.

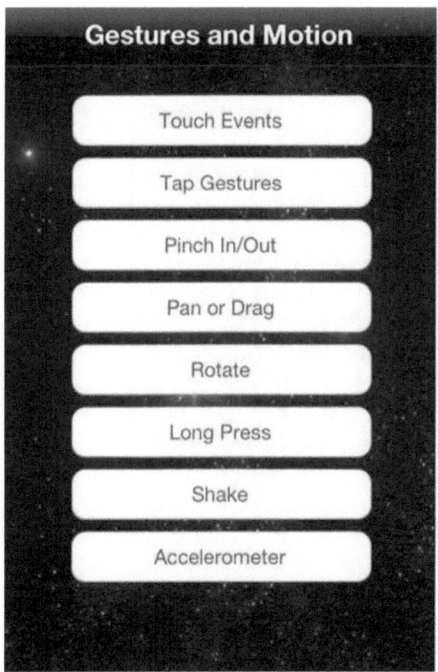

Figure 8. User input example list with one example

On the left of Figure 8 we see eight buttons, each labeled with a type of user input. In Chapter 8, we will systematically go through each input type and build an example that shows how it works. On the right side of Figure 8, we see the example for Touch Events. We use the classes we defined in Chapters 6 and 7 to build each example, so we understand not just the gestures in general, but how they relate to working with our game classes.

Chapter 9

The great thing about building games these days is that a network connection is ubiquitous. This network connection allows our game to reach outside of the device it is running on and use the vast resources of the Internet. For games, this means social networking. In Chapter 9, we learn about Apples Game Center and related GameKit library. We also look at Twitter and Facebook. Figure 9 shows an example of using these services.

Figure 9. In-game social media

In Figure 9, we see on the left the Achievement view from Game Center for our example app, Belt Commander. On the right we see a tweet with default text. (Facebook is not shown.)

The code that enables these services is implemented in project for our sample game Belt Commander. In this way, we learn about these services with the context of applying them to a real game.

In addition to implementing the code required to enable these services, we look at how each of these services must be enabled outside of our game code. For example, to use Game Center features in your application, you have to enable that service in iTunes Connect. We also look at what is required outside of the code to get Facebook and Twitter working.

Chapter 10

The last chapter in this book is dedicated to monetizing your game. We look at the details of including in-app purchases in your application. We walk through the details of setting up in-app purchases in iTunes Connect as well as the code within the app. Ideally, we want the user to be able to easily reach the dialog shown in Figure 10.

Figure 10. *A successful purchase makes the developer some money.*

Chapter 11

In Chapter 11, we review all you have learned by walking through the sample game, Belt Commander. We look at how the application is organized in terms of views and the navigation between them. We look at the specifics of extending GameController to define the logic that governs how the game is played and the interactions between the actors. We also look at the code that gives each actor in the game its unique behavior. In short we show the game Belt Commander is implemented, as shown in Figure 11.

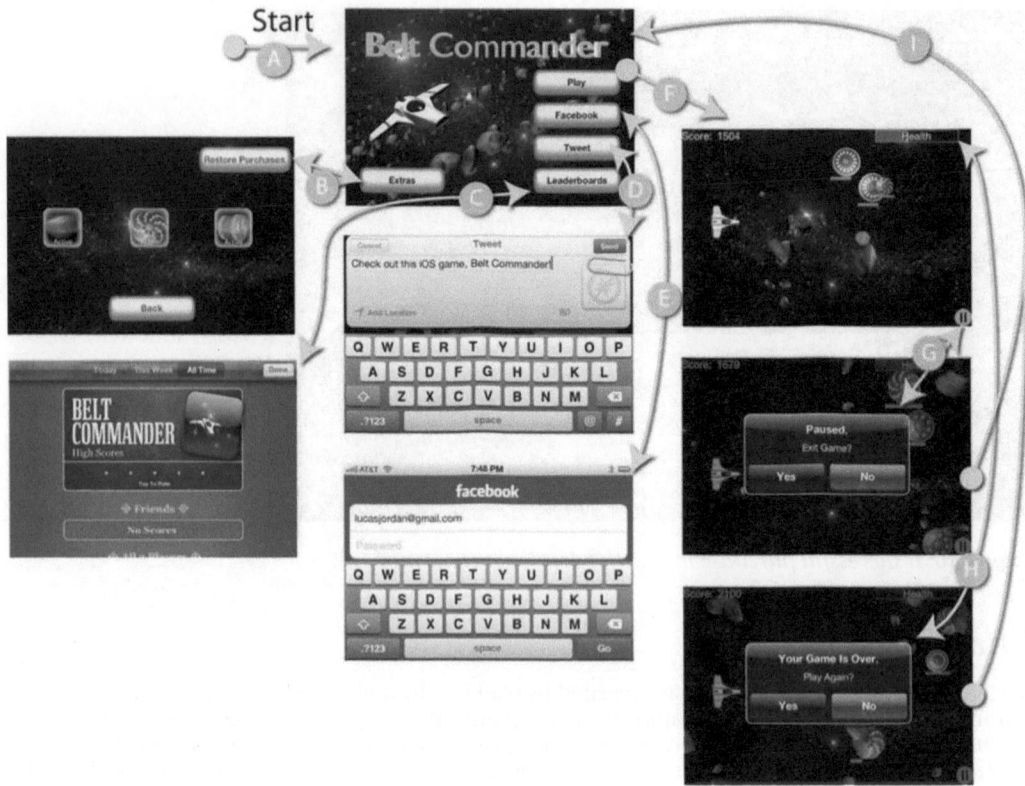

Figure 10. *An action shot in Belt Commander*

In Figure 10, we see all of the main screens of the game Belt Commander. By the end of this chapter, you will know how each works. At that point, you will have learned all that you need to know to create compelling games on your own.

Appendix A

Appendix A takes a break from Belt Commander and looks at some techniques that can be used to make graphics for games. We also explore some of the reasoning that goes into figuring out what size you final art assets should be. Figure 12 shows one of the diagrams from Appendix A.

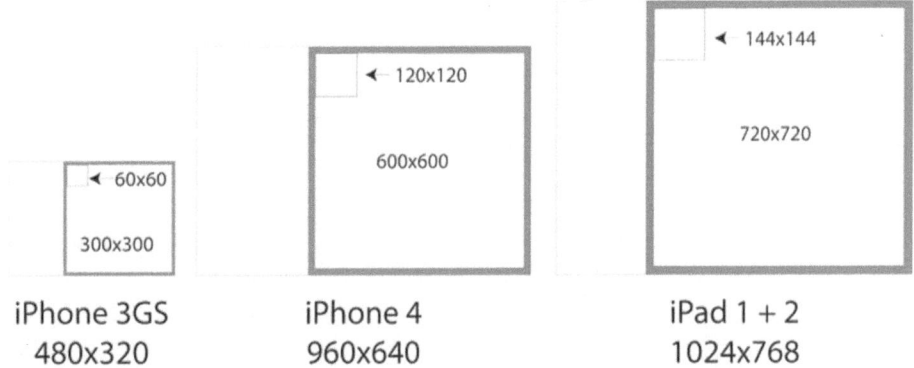

Figure 12. *Comparative graphic sizes*

In Figure 12, we see the three screen sizes for an iOS devices. This diagram shows how to decide what sizes your images should be when you implement your game. Image size is important because of memory usage and image quality when displayed on the screen.

Appendix A also spells out how to create and identify the images that support your app. For example, we look at creating icon images that support all three iOS devices. Appendix A also takes a moment to talk about style in a game. This discussion describes the importance of consistency over quality when it comes to art.

Chapter 1

A Simple First Game

In this chapter, we are going to build a very simple game of Rock, Paper, Scissors. We will use the Storyboard feature of Xcode to create an application with two views and the navigation between them.

Included with this book are sample Xcode projects; all of the code examples are taken directly from these projects. In this way, you can follow along with each one in Xcode. I used version 4.2 of Xcode when creating the projects for this book. The project that accompanies this chapter is called Sample 1. The project is a very simple game in which we use Storyboard to create two scenes. The first scene is the starting view, and the second scene is where the user can play the Rock, Paper, Scissors game. The second scene is where you will add a UIView and specify the class as RockPapaerScissorView.

We will walk through each of these steps, but first let's take a quick look at our game, shown in Figure 1–1.

On the left of Figure 1–1 we see the starting view. It just has a simple title and a Play button. When the user clicks the Play button, he is transitioned to the second view, shown on the right of the figure. In this view, the user can play Rock, Paper, Scissors. If the user wishes to return to the starting view, or home screen, he can press the Back button. This simple game is composed of a Storyboard layout in Xcode and a custom class that implements the game.

Let's take a look at how I created this game and at some ways you can customize a project.

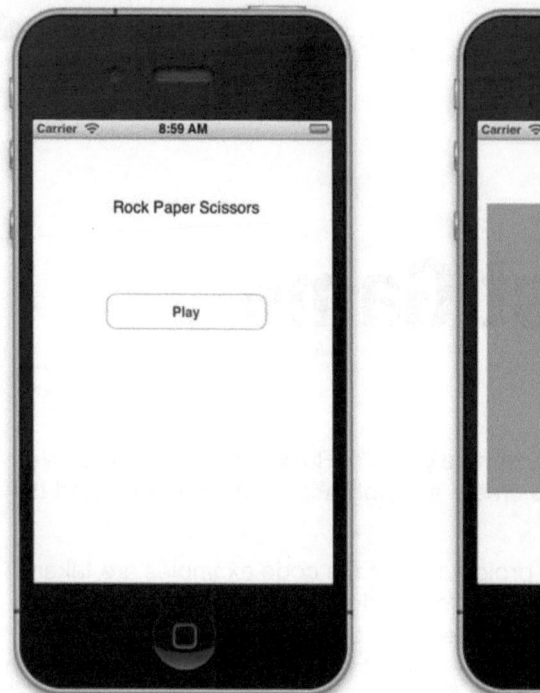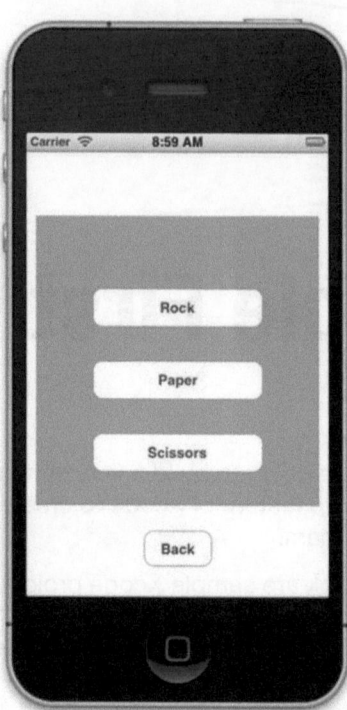

Figure 1–1. *The two views of our first game: Sample 1*

Creating a Project in Xcode: Sample 1

Creating this game only involves a few steps, which we'll walk through as an introduction to Xcode.

Start by launching Xcode. From the File menu, select New Project. You will see a screen showing the types of projects you can create with Xcode (See Figure 1–2).

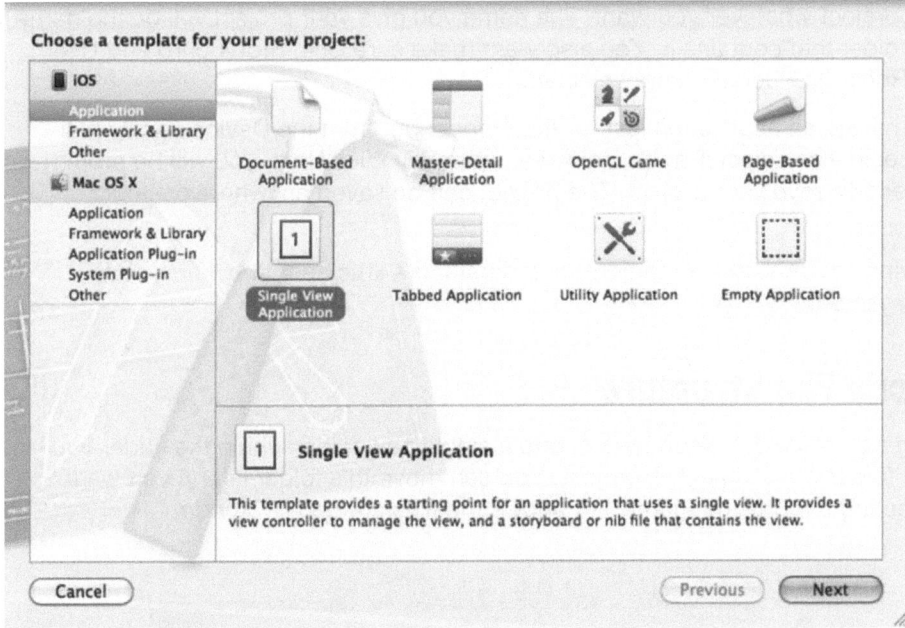

Figure 1–2. *Project templates in Xcode*

For this project, select the template Single View Application. Click Next, and you will be prompted to name the project, as shown in Figure 1–3.

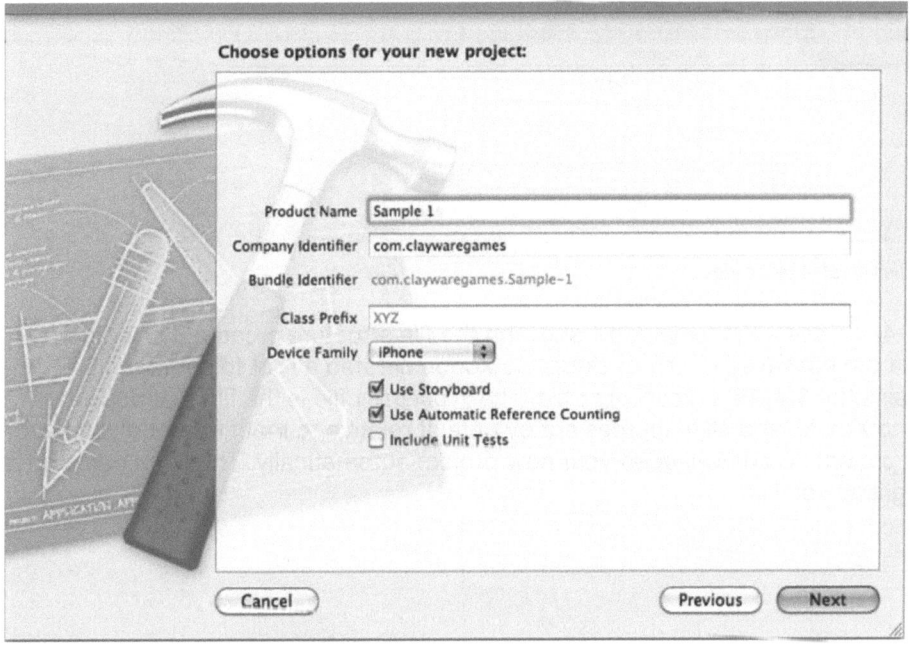

Figure 1–3. *Naming an Xcode project*

Name your project whatever you want. The name you give your project will be the name of the root folder that contains it. You also want make sure Use Storyboard and Use Automatic Reference Counting are selected.

We will be making an application just for the iPhone, but from the Device Family pull-down you could also select iPad or Universal. After you click Next, you will be prompted to pick a place to save your project. The project can be saved anywhere on your computer.

Before moving on, let's take a moment to understand a little about how an Xcode project is organized.

A Project's File Structure

After saving a new project, Xcode will create a single new folder within the folder you select. This folder will contain the project. You can move this folder later if you want without affecting the project. Figure 1–4 shows the files created by Xcode.

Figure 1–4. *Files created by Xcode*

In Figure 1–4, we see a Finder window showing the file structure created. I selected that I wanted the project saved on my desktop, so Xcode created a root folder name Sample 1 that contains the Sample 1.xcodeproj file. The xcodeproj file is the file that describes the project to Xcode, and all resources are by default relative to that file. Once you have saved your project, Xcode will open your new project automatically. Then you can start customizing it as you like.

Customizing Your Project

We have looked at how to create a project. Now you are going to learn a little about working with Xcode to customize your project before moving on to adding a new UIView that implements the game.

Arranging Xcode Views to Make Life Easier

Once you have a new project created, you can start customizing it. You should have Xcode open with your new project at this point. Go ahead and click the MainStoryboard.storyboard file found on the left so your project looks like Figure 1–5.

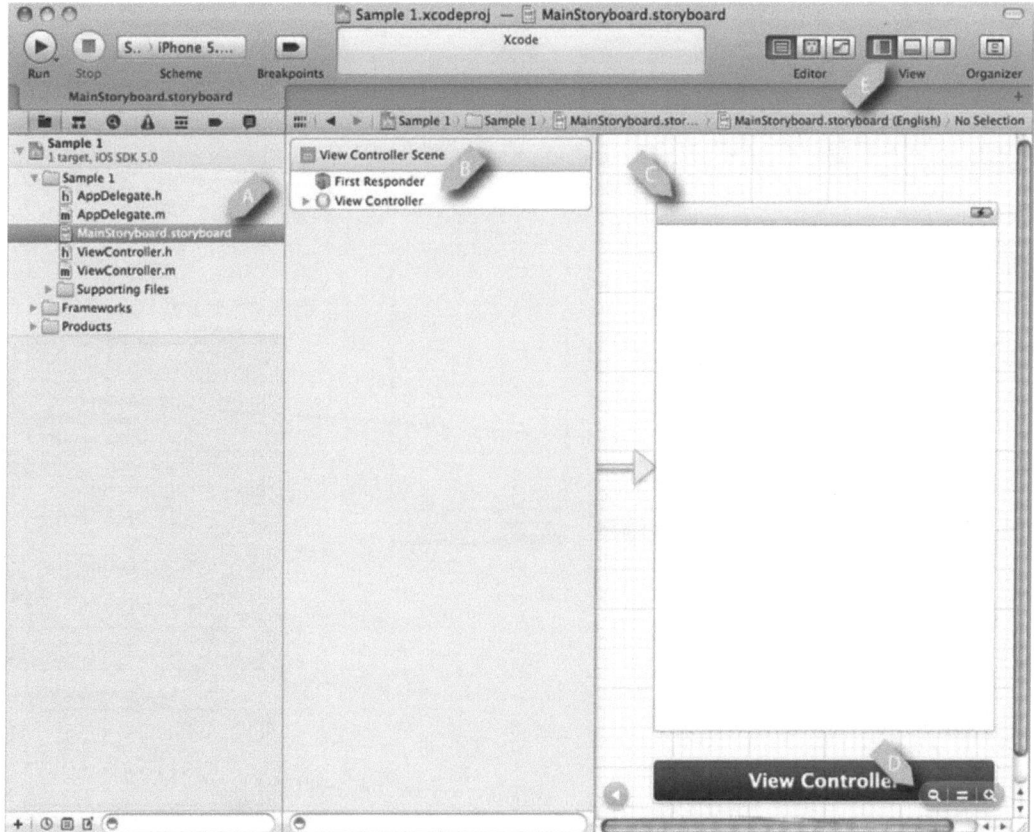

Figure 1–5. *MainStoryboard.storyboard before customization*

In Figure 1–5, we see the file MainStoryboard.storyboard selected (item A). This file is used to describe multiple views and the navigation relationships between them. It shows the selected storyboard file and describes the content of the right side of the screen. In item B, we see an item called View Controller. This is the controller for the view described in item C. We will look at how Xcode works in more detail over the next few chapters. I do want to point out the controls labeled item D. They are used to zoom in and out of a storyboard view, and are critical to successfully navigating your way around. Additionally, the buttons in item E are used to control which of the main panels are visible in Xcode. Go ahead and play around with those buttons.

Next, let's look at how to add a new view.

Adding a New View

Once you have had a chance to play a little with the different view setups available in Xcode, you can move on and add a new view to your project. Arrange Xcode so the right-most panel is visible, and hide the left-most panel if you want. Xcode should look something like Figure 1–6.

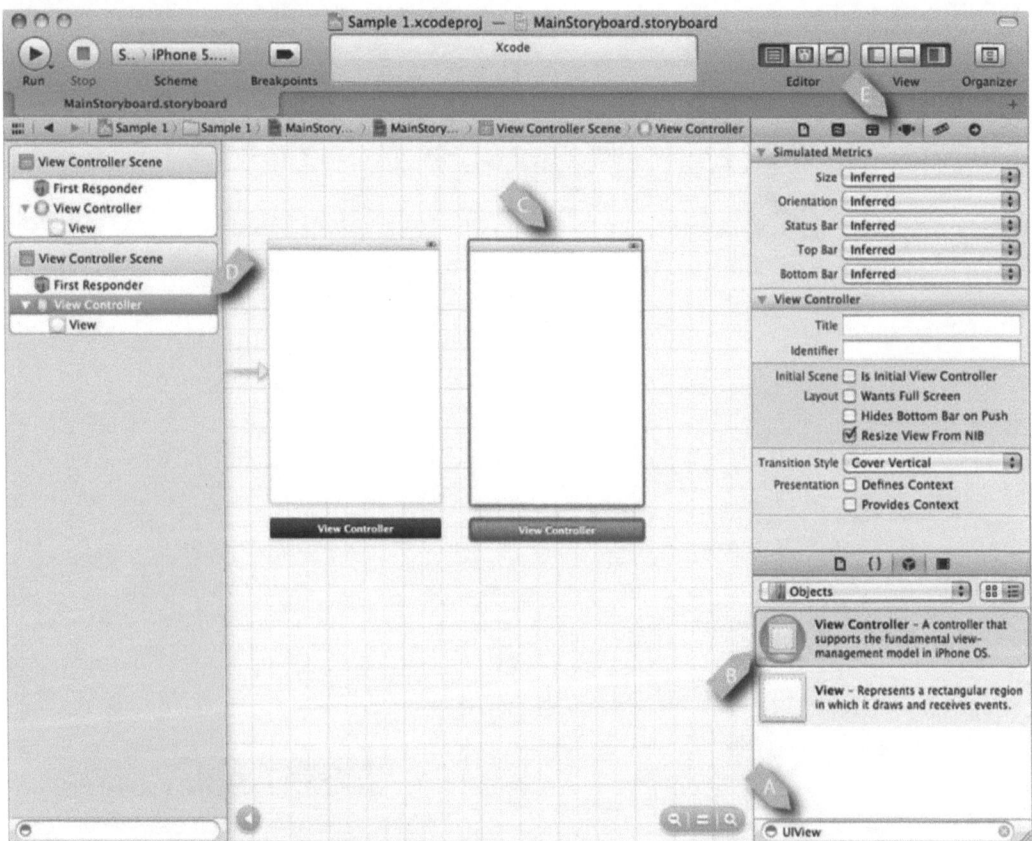

Figure 1–6. *Storyboard with second view*

In Figure 1–6, we see that we have added a second view to the storyboard. Like any good Apple desktop application, most of the work is done through dragging and dropping. To add the second view we enter the words **UIView** into the bottom right text field, item A. This filters the list so we can drag the icon labeled item B on the work area in the center. Click on the new view so it is selected (see item C), which we can see correlates to the selected icon in item D. Item E shows the properties for the selected item.

Now that we have a new view in the project, we want to set up a way to navigate between our views.

Simple Navigation

We now want to create some buttons that enable us to navigate from one view to the other. The first step is to add the buttons, and then to configure the navigation. Figure 1–7 shows these views being wired up for navigation.

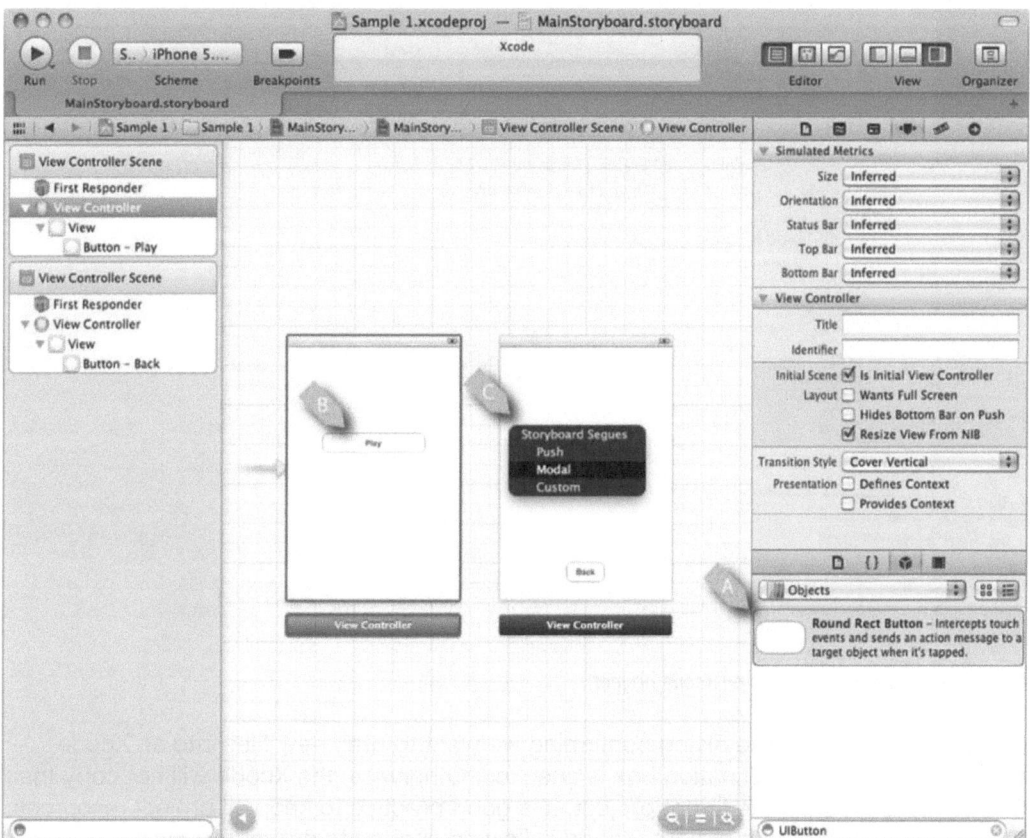

Figure 1–7. *Storyboard with navigation*

In Figure 1–7, we see that we have dragged a UIButton from the library item A onto each of the views. We gave the UIButton on the left the label Play, and the UIButton on the right the label Back. To make the Play button navigate to the view on the right, we right-drag from the Play button (item B) to the view on the right and release at item C. When we do this, a context dialog pops up, allowing us to select which type of transition we want. I selected Model. We can repeat the process for the Back button: right-drag it to the view on the left and select which transition you want for the return trip. You can run the application at this point and navigate between these two views. In order to make it a game, though, we need to include the Rock, Paper, Scissors view and buttons.

Adding the Rock, Paper, Scissors View

To add the Rock, Paper, Scissors view, we need to include a class from the sample code in the project you are building. The easiest way to do this is to open the sample project and drag the files RockPaperScissorsView.h and RockPaperScissorsView.m from the sample project to the new project. Figure 1–8 shows the dialog that pops up when you drag files into an Xcode project.

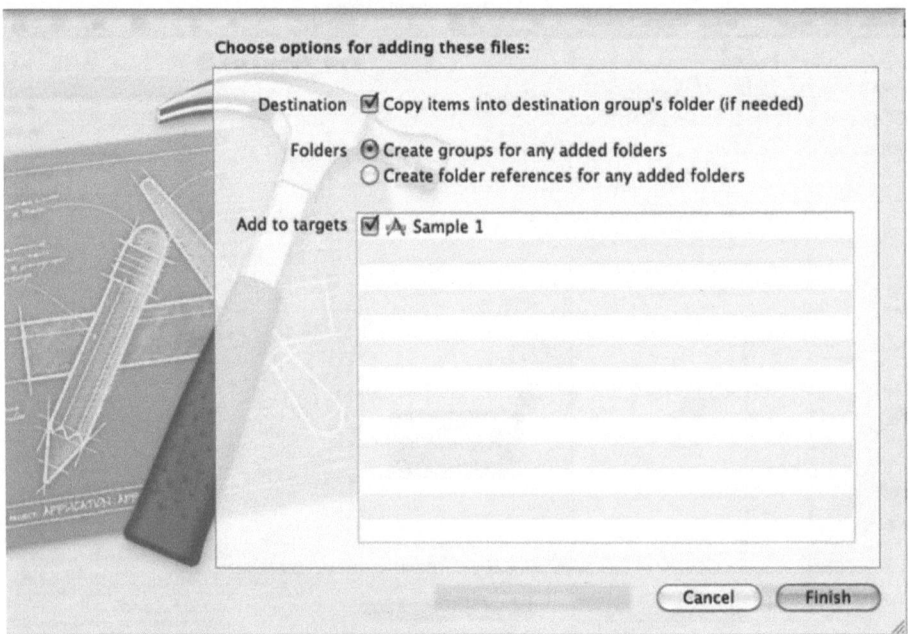

Figure 1–8. *Dragging files into an Xcode project*

In Listing 1–8, we see the dialog confirming we want to drag new files into an Xcode project. Be sure the Destination box is checked. Otherwise, the Xcode will not copy the files to the location of the target project. It is good practice to keep all project resources under the root folder of a project. Xcode is flexible enough to not require that you do this, but I have been burned too many times by this flexibility. Anyway, now that we have the required class in our project let's wire up our interface to include it.

Customizing a UIView

The last step in creating a simple application is to create a new UIView in our interface that is of the class RockPaperScissorsView. Figure 1–9 shows how this is done.

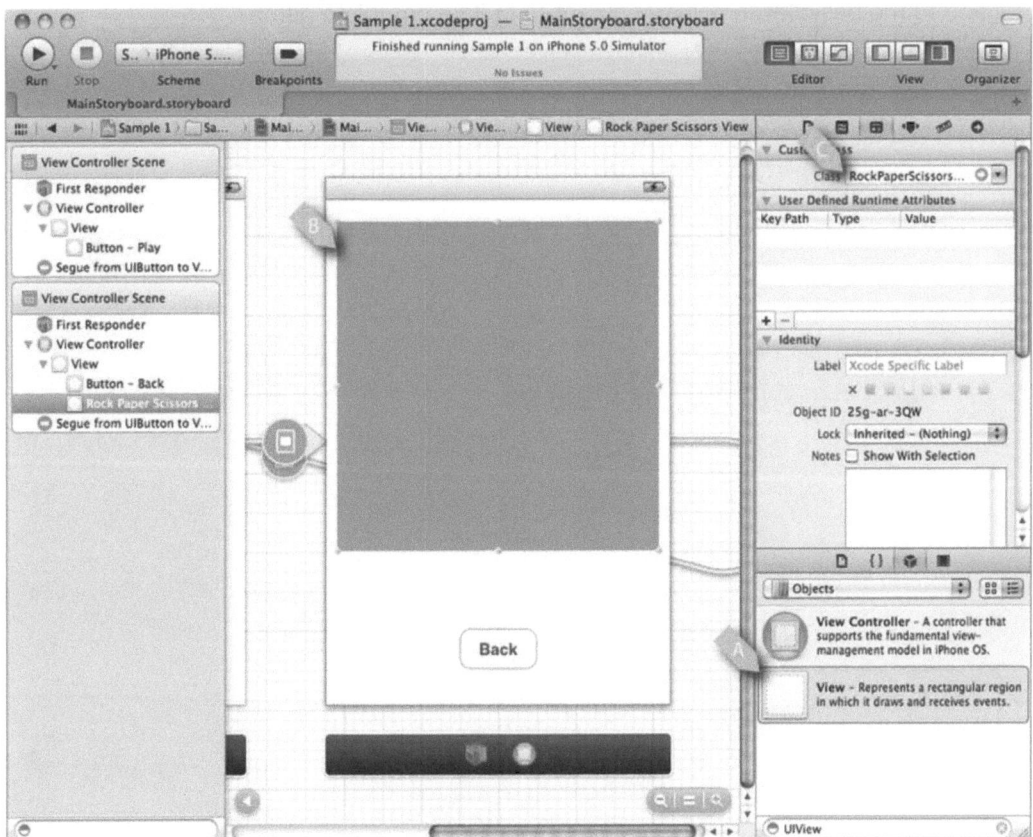

Figure 1–9. *A customized UIView*

In Figure 1–9, we see a UIView added to the view on the right. We did this by dragging the icon from item A onto the storyboard in item B. After adjusting the size of the new UIView, we set its class to be RockPaperScissorsView, as shown in item C. At this point, we are technically done. You have created our first game! Obviously, we have not looked at the implementation of RockPeperScissorsView, which is discussed on the next chapter.

The rest of this book will use Sample 1 as a starting place. We will learn many new techniques for customizing a simple app to make a truly complete game.

Summary

In this chapter, you have taken a quick tour through Xcode, learning how to create a project with it and build a simple navigation using Storyboard. The chapters that follow will add to the basic lessons learned here to build a complete game.

Chapter 2

Setting Up Your Game Project

Like all software projects, iOS game development benefits from starting on good footing. In this chapter, we will discuss setting up a new Xcode project that is a suitable starting point for many games. This will include creating a project that can be used for the deployment on the iPhone and the iPad, handling both landscape and portrait orientations.

We look at how an iOS application is initialized and where we can start customizing behavior to match our expectations of how the application should perform. We will also explore how user interface (UI) elements are created and modified in an iOS application, paying special attention to managing different devices and orientations.

The game we create in this chapter will be very much like the simple example from Chapter 1—in fact, it will play exactly the same. But we will be building a foundation for future chapters while practicing some key techniques, such as working with `UIViewControllers` and Interface Builder.

We will explore how an iOS application is put together, and explain the key classes. We'll also create new UI elements and learn how to customize them with Interface Builder, and we will explore using the MVC pattern to create flexible, reusable code elements. At the end of this chapter, we will have created the Rock, Paper, Scissors application shown in Figure 2–1.

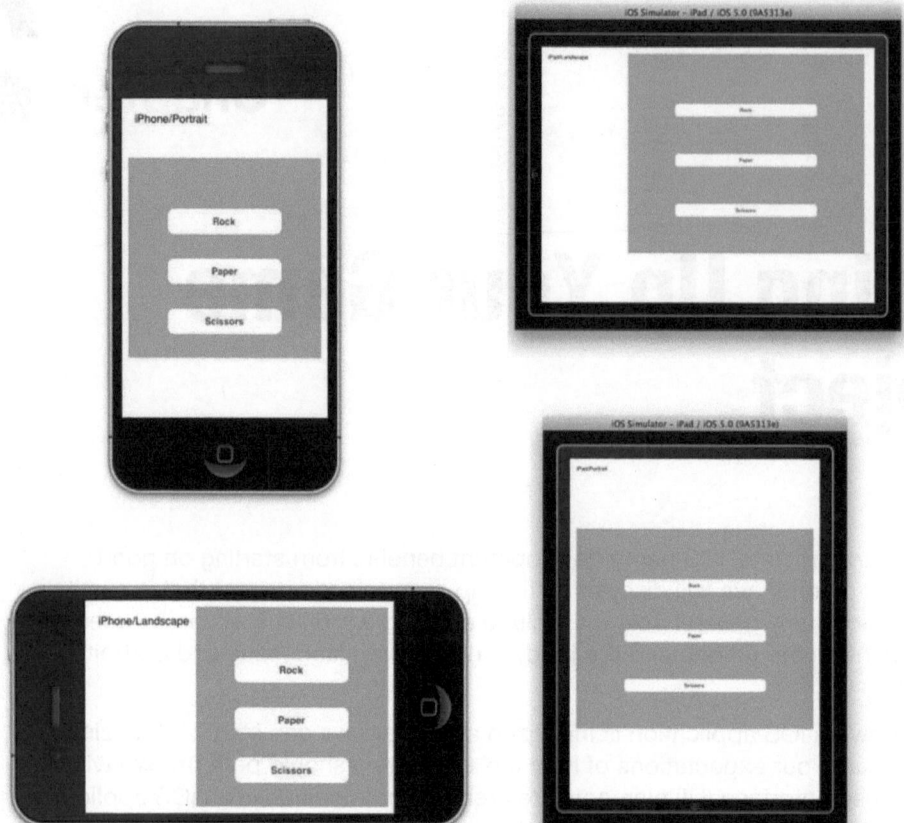

Figure 2–1. *An application design to work on the iPhone and iPad in all orientations*

Figure 2–1 shows the application running on an iPhone and iPad simulator. This is a so-called *universal* application: it can run on both devices and would be presented in the App Store as such. Unless there are specific business reasons for writing an app that only works on the iPhone or the iPad, it makes a lot of sense to make your app universal. It will save you time down the road, even if you only intend to release your app on one of the devices to start.

Our sample application is so simple that it may be difficult to see the differences between the four states presented in Figure 2–1. On the upper left, where the iPhone is in portrait orientation, the position of the gray area is laid out differently from the iPhone in landscape at the bottom left. The layout of the text is also different. The same goes for the application when it is running on the iPad in landscape vs. portrait. Let's get going and understand how we can set up a project to accommodate these different devices and orientations.

Creating Your Game Project

To get things started with our sample game, we first have to create a new project in Xcode.

Create a new project by selecting File > New > New Project…. This will open a wizard that allows you to select which type of project you want, as shown in Figure 2–2.

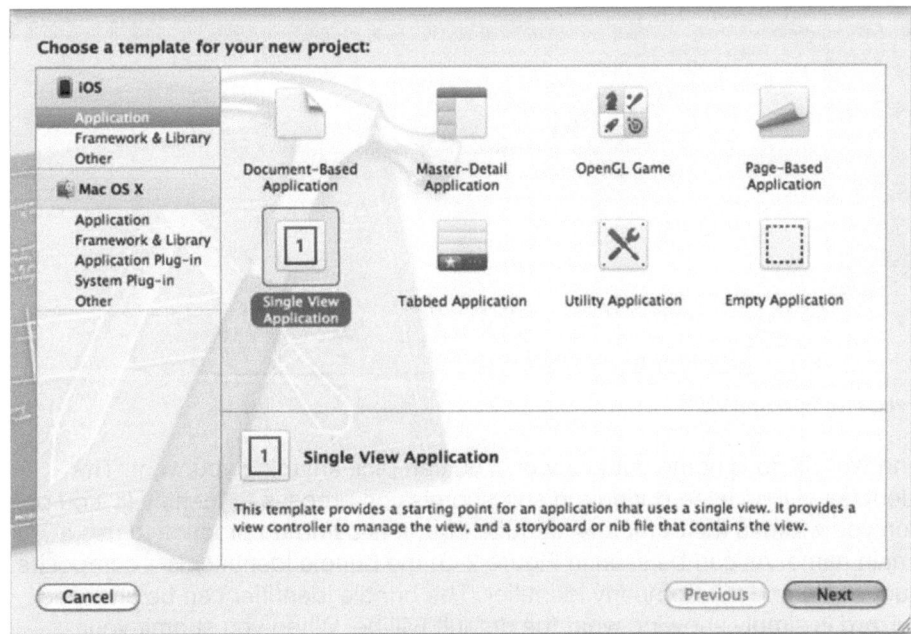

Figure 2–2. *Creating a new window-based application*

On the left side of Figure 2–2, we see we have selected Application from the iOS section. On the right are the available project types for creating iOS applications. The choices presented here help developers by giving them a reasonable starting place for their applications. This is particularly helpful for developers new to iOS, because these templates get you started for a number of common application navigation styles. We are going to pick a Single View Application, because we only require a very minimal starting point, and the Single View Application provides good support for universal applications. After clicking Next, we see the options shown in Figure 2–3.

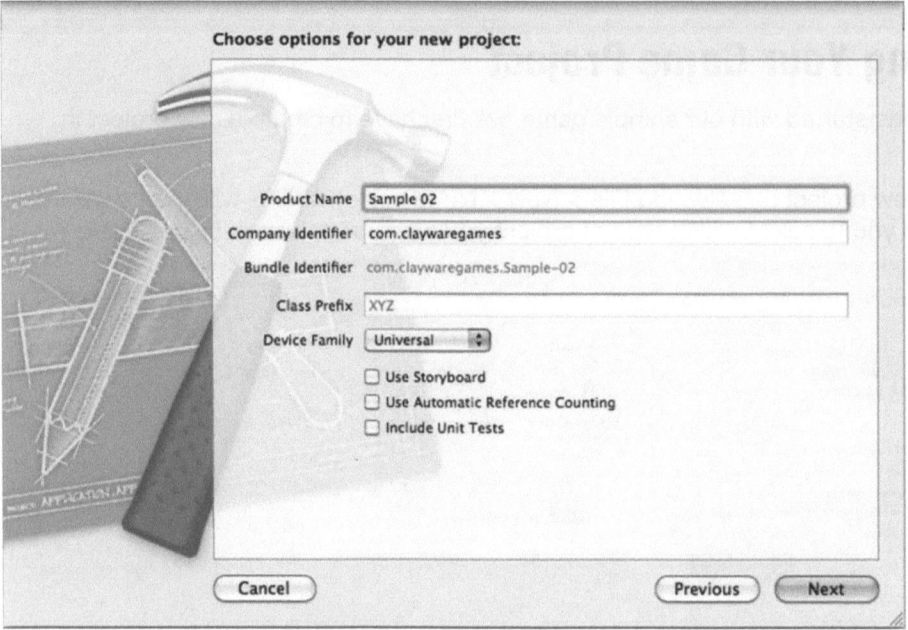

Figure 2–3. *Details for the new project*

The first thing we will do is name our product. You can pick anything you want. The company identifier will be used during the app submission process to identify it. You can put any value you want as the company identifier, but it is common practice to use a reverse domain name. As can be seen in Figure 2–3, the bundle identifier is a composite of the product name and the company identifier. The bundle identifier can be changed later—the wizard is simply showing what the default will be. When you submit your game to the App Store, the bundle identifier is used to indicate which application you are uploading.

By selecting Universal from the Device Family list, you are telling Xcode to create a project that is ready to run on both the iPhone and iPad. For this example, we will not be using Storyboard ot Automatic Reference Counting. Similarly, we won't be creating any unit test, so the Include Unit Tests option should be unchecked as well. Clicking Next prompts you to save the project. Xcode will create a new folder in the selected directory, so you don't have to manually create a folder if you don't want to. When the new project is saved, you will see something similar to Figure 2–4.

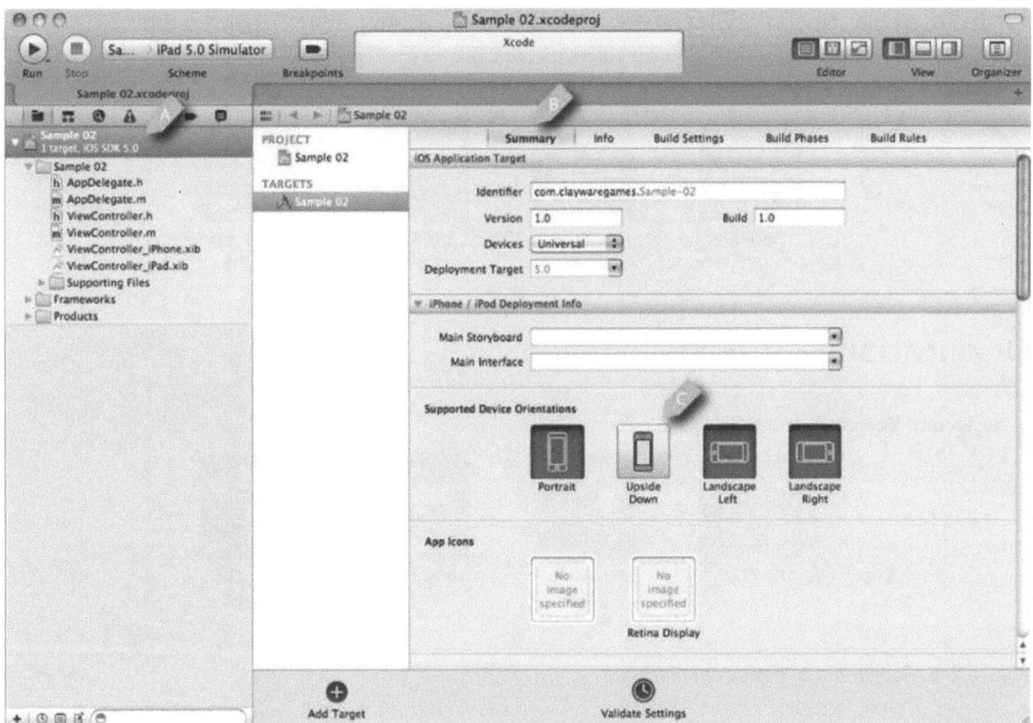

Figure 2–4. *A newly created project*

On the left side of Figure 2–4, there is a tree containing the elements of the project with the root most element selected (A). On the right, the Summary tab is selected (B). From the Summary tab, we want to select the supported device orientations (C). To support both orientations on each device, click the Upside Down button. Scroll down and make sure the all of the orientations for the iPad are depressed as well. Figure 2–5 shows the correct settings. Now that the project is created, it is time to start customizing it to fit our needs.

Orientations for iPhones

Orientations for iPads

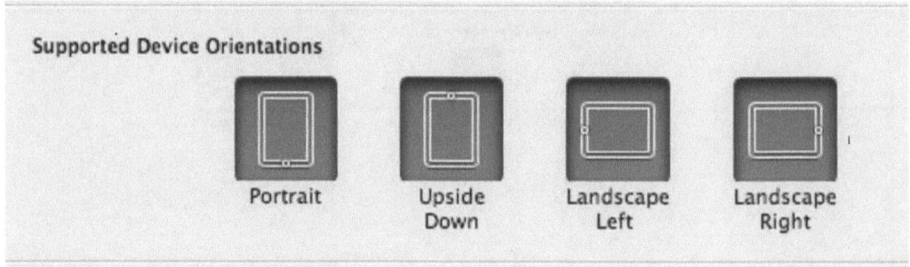

Figure 2–5. *Supporting all device orientations*

Customizing a Universal Application

In order to understand the customizations we will make to this project, it is good to understand where we start. The empty project is runnable as is, although it is obviously not very interesting. Take a moment and run the application. It will look something like the app running in Figure 2–6. Next we will be removing the status bar and exploring the best place to start adding custom code to our project.

Figure 2–6. *A fresh universal iOS application*

In Figure 2–6, we see a new universal app running on the iPhone simulator. To run the application on the iPad simulator, select iPad Simulator from the Scheme pull-down menu to the right of the Stop button in Xcode. As we can see, the app is empty; there is only a gray background. Another thing to note is that the status bar at the top of the application is displayed. Although there are plenty of good reasons to include the status bar in an application, many games may wish to remove the status bar in order to create a more immersive experience. To remove the status bar, click on the root element in the Project Navigator (the tree on the left in Xcode). Select the target and then click the Info tab on the right. Figure 2–7 shows the correct view.

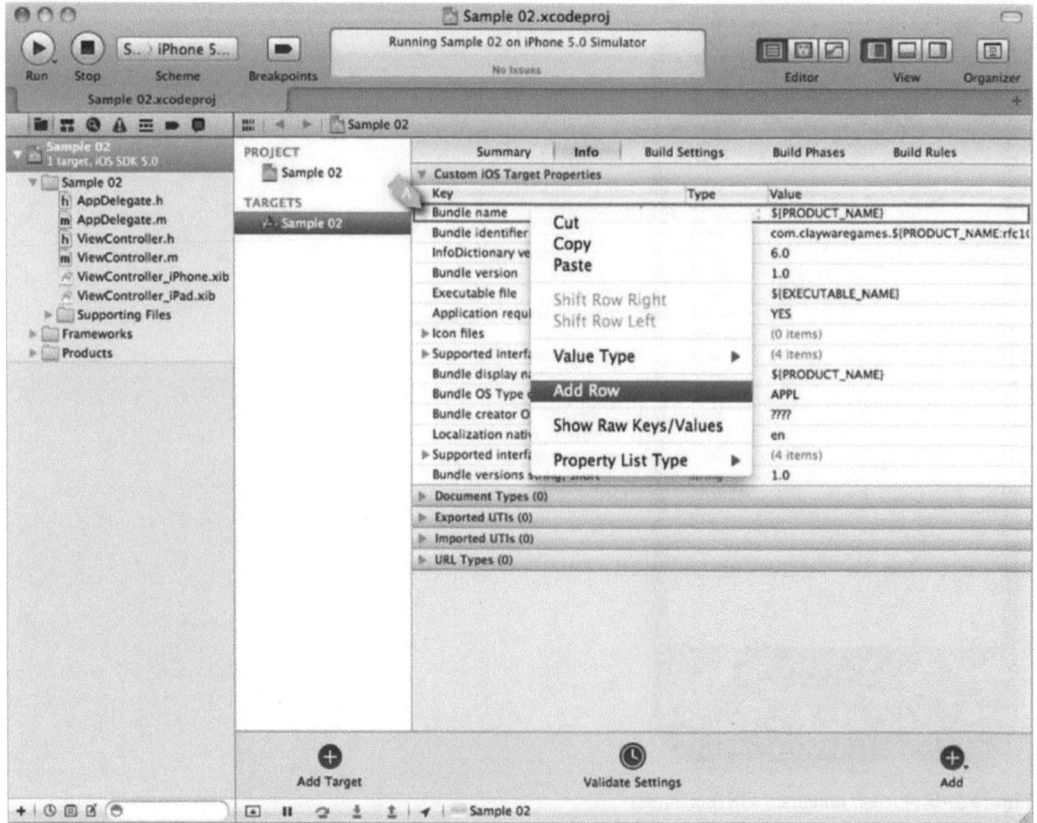

Figure 2–7. *Configuring the status bar*

Once you see the view shown in Figure 2–7, right-click on the top-most element (A) and select Add Row. This will add a new element to the list of items. You will want to set the key value to "Status bar is initially hidden" and the value to "Yes". What you are doing here is editing the plist file found under the group Supporting Files. Xcode is simply providing you with a nice interface for editing this configuration file.

> **TIP:** Navigate to the file ending with info.plist under the group Supporting Files. Right-click on it and select Open As >Source Code. This will show you the contents of the plist file source. Note that the key values are actually stored as constants and not as the human readable text used in the Info editor.

When we run the application again, the status bar is removed, as shown in Figure 2–8.

Figure 2–8. *The default application without the status bar*

Now that we have explored a simple way to start customizing our application, it is time to investigate just how an iOS application is put together, so we can make smart decisions about adding our own functionality.

How an iOS Application Initializes

As we know, iOS applications are written largely in Objective C, which is a super set of C. Xcode does a good job of hiding the details of how an application is built, but under the covers we know it is using LLVM and other common Unix tools to do the real work. Because we know that our project is ultimately a C application, we would expect the starting point for our application to be a C main function. In fact, if you look under the Supporting Files group, you will find the main.m, as shown in Listing 2–1.

Listing 2–1. *main.m*

```
#import <UIKit/UIKit.h>

#import "AppDelegate.h"

int main(int argc, char *argv[])
{
    @autoreleasepool {
```

```
        return UIApplicationMain(argc, argv, nil, NSStringFromClass([AppDelegate
class]));
    }
}
```

The main function in Listing 2–1 is a C function, but you will notice that the body of the function is clearly Objective C syntax. An @autoreleasepool wraps a call to UIApplicationMain. The @autoreleasepool sets up the memory management context for this application, we don't really need to worry about that. The function UIApplicationMain does a lot of handy stuff for you—it initializes your application by looking at your info.plist file, sets up an event loop, and generally gets things going in the right way. In fact, I have never had any reason to modify this function, because iOS provides a concise place to start adding your startup code. The best place to start adding initialization code is the task application:didFinishLaunchingWithOptions: found in the implementation file of your app delegate class. In this example, the app delegate class is called AppDelegate. Listing 2–2 shows the implementation of the application:didFinishLaunchingWithOptions: task from AppDelegate.m.

Listing 2–2. *application:didFinishLaunchingWithOptions:*

```
- (BOOL)application:(UIApplication *)application
didFinishLaunchingWithOptions:(NSDictionary *)launchOptions
{
    self.window = [[[UIWindow alloc] initWithFrame:[[UIScreen mainScreen] bounds]]
autorelease];
    // Override point for customization after application launch.
    if ([[UIDevice currentDevice] userInterfaceIdiom] == UIUserInterfaceIdiomPhone) {
        self.viewController = [[[ViewController_iPhone alloc]
initWithNibName:@"ViewController_iPhone" bundle:nil] autorelease];
    } else {
        self.viewController = [[[ViewController_iPad alloc]
initWithNibName:@"ViewController_iPad" bundle:nil] autorelease];
    }
    self.window.rootViewController = self.viewController;
    [self.window makeKeyAndVisible];
    return YES;
}
```

When an application is fully initialized and ready to start running the code that comprises it, the task application:didFinishLaunchingWithOptions: is called, as seen in Listing 2–2. This task, besides having an incredibly long name, takes an instance of UIApplication and NSDictionary as arguments. The UIApplication is an object that represents the state of the running application. The instance of AppDelegate that receives this message is the delegate to the UIApplication object. This means that we modify the application's behavior by implementing tasks in AppDelegate. The protocol UIApplicationDelegate, which AppDelegate implements, defines the tasks that will be called on the application's behalf.

If we take a look at the implementation of application:didFinishLaunchingWithOptions: we start by creating a new UIWindow with a size equal to the size of the screen. Once the UIWindow is instantiated we need to create a UIViewController that will manage the UI of our application. A new Xcode project, like the one we created, starts out with a single ViewController class to manage our UI. We are going to create device specific subclasses of our ViewController, so that we have a place to put device specific code. We will look at the process of creating these subclasses shortly. To make sure that we instantiate the correct ViewController subclass we need to add the bold code in Listing 2–2.

Once our ViewController is created we set it as the rootViewController of window and call makeKeyAndVisible. The window object is an instance of UIWindow and is the root graphical component of your application. The task makeKeyAndVisible essentially displays the window. Any custom components we add to our application will be subviews of this window object.

If your application requires that configuration be done before the UI is loaded, you put that initialization code before the call to makeKeyAndVisible. This might include reading an app-specific configuration file, initializing a database, setting up the location service, or any number of other things.

Next we will take a general look at how a UI is organized in an iOS application before looking at the details of working with Interface Builder.

Understanding UIViewControllers

iOS development and the associated libraries make heavy use of the Model View Controller (MVC) pattern. In general, MVC is a strategy for separating the presentation (View), data (Model), and business logic (Controller). In specific terms, the model is simply data, like a Person class or an Address. The view is responsible for rendering the data to the screen. In iOS development, that means a subclass of UIView. iOS provides a special class to act as the controller for a UIView, which is aptly named UIViewController.

UIViewController has two key characteristics: it is often associated with an XIB file and it has a property called "view" that is of type UIView. By creating a subclass of UIViewController, we can also make an XIB file with the same name as the class. By default, when you instantiate a UIViewController subclass, it will load a XIB with the same name. The root UIView in the XIB will be wired up as the view property of UIViewController.

Besides providing a clean separation between the UI layout and the logic that drives it, iOS provides a number of UIViewController subclasses that expect to work with other UIViewControllers instead of UIViews. An example of this is the UINavigationController that implements the type of navigation found in the Settings app. In code, when you want to advance to the next view, you pass a UIViewController and not a UIView, though it is the view property of the UIViewController that is displayed on the screen.

Admittedly, for our example application in this chapter it does not make a lot of difference if we use a UIViewController. In Chapter 1, we extended UIView when we created the RockPaperScissorsView class and it worked fine. However, understanding how UIViewControllers and their views work together will make life easier in Chapter 3, where we explore a game's application life cycle.

Using the RockPaperScissorsView from Chapter 1, let's take a look at how this functionality would look if it were implemented as a UIViewController. Listing 2–3 shows the file RockPaperScissorsController.h.

Listing 2–3. *RockPaperScissorsController.h*

```
@interface RockPaperScissorsController : UIViewController {
    UIView* buttonView;
        UIButton* rockButton;
        UIButton* paperButton;
        UIButton* scissersButton;

        UIView* resultView;
        UILabel* resultLabel;
        UIButton* continueButton;

    BOOL isSetup;
}
-(void)setup:(CGSize)size;
-(void)userSelected:(id)sender;
-(void)continueGame:(id)sender;

-(NSString*)getLostTo:(NSString*)selection;
-(NSString*)getWonTo:(NSString*)selection;
@end
```

In listing 2–3, we see the class RockPaperScissorsController extends UIViewController. Among other things, this means that RockPaperScissorsController has a property called "view" that will be the root UIView for this controller. Like RockPaperScissorsView, we will have other UIViews that are subviews of the root view, such as the buttons for selecting your choice. Although these buttons could conceivably have their own UIViewControllers, there comes a point when it makes sense to allow a UIViewController to manage the all of the UIViews with which it is concerned. There are very few changes required on the implementation side of things to complete this transition from UIView to UIViewController. Basically, wherever we used the keyword "self," we simply say self.view instead. Listing 2–4 shows the required changes.

Listing 2–4. *RockPaperScissorsController.m (Partial)*

```
-(void)setup:(CGSize)size{

    if (!isSetup){
        isSetup = true;

        srand(time(NULL));

        buttonView = [[UIView alloc] initWithFrame:CGRectMake(0, 0, size.width,
size.height)];
        [buttonView setBackgroundColor:[UIColor lightGrayColor]];
        [self.view addSubview:buttonView];

        float sixtyPercent = size.width * .6;
        float twentyPercent = size.width * .2;
        float twentFivePercent = size.height/4;
        float thirtyThreePercent = size.height/3;
```

```
        rockButton = [UIButton buttonWithType:UIButtonTypeRoundedRect];
        [rockButton setFrame:CGRectMake(twentyPercent, twentFivePercent, sixtyPercent,
40)];
        [rockButton setTitle:@"Rock" forState:UIControlStateNormal];
        [rockButton addTarget:self action:@selector(userSelected:)
forControlEvents:UIControlEventTouchUpInside];

        paperButton = [UIButton buttonWithType:UIButtonTypeRoundedRect];
        [paperButton setFrame:CGRectMake(twentyPercent, twentFivePercent*2,
sixtyPercent, 40)];
        [paperButton setTitle:@"Paper" forState:UIControlStateNormal];
        [paperButton addTarget:self action:@selector(userSelected:)
forControlEvents:UIControlEventTouchUpInside];

        scissersButton = [UIButton buttonWithType:UIButtonTypeRoundedRect];
        [scissersButton setFrame:CGRectMake(twentyPercent, twentFivePercent*3,
sixtyPercent, 40)];
        [scissersButton setTitle:@"Scissers" forState:UIControlStateNormal];
        [scissersButton addTarget:self action:@selector(userSelected:)
forControlEvents:UIControlEventTouchUpInside];

        [buttonView addSubview:rockButton];
        [buttonView addSubview:paperButton];
        [buttonView addSubview:scissersButton];


        resultView = [[UIView alloc] initWithFrame:CGRectMake(0, 0, size.width,
size.height)];
        [resultView setBackgroundColor:[UIColor lightGrayColor]];

        resultLabel = [[UILabel new] initWithFrame:CGRectMake(twentyPercent,
thirtyThreePercent, sixtyPercent, 40)];
        [resultLabel setAdjustsFontSizeToFitWidth:YES];
        [resultView addSubview:resultLabel];

        continueButton = [UIButton buttonWithType:UIButtonTypeRoundedRect];
        [continueButton setFrame:CGRectMake(twentyPercent, thirtyThreePercent*2,
sixtyPercent, 40)];
        [continueButton setTitle:@"Continue" forState:UIControlStateNormal];
        [continueButton addTarget:self action:@selector(continueGame:)
forControlEvents:UIControlEventTouchUpInside];
        [resultView addSubview:continueButton];

    }
}
-(void)userSelected:(id)sender{
        int result = random()%3;

        UIButton* selectedButton = (UIButton*)sender;
        NSString* selection = [[selectedButton titleLabel] text];

        NSString* resultText;
        if (result == 0){//lost
                NSString* computerSelection = [self getLostTo:selection];
                resultText = [@"Lost, iOS selected " stringByAppendingString:
computerSelection];
        } else if (result == 1) {//tie
```

```
                        resultText = [@"Tie, iOS selected " stringByAppendingString: selection];
            } else {//win
                        NSString* computerSelection = [self getWonTo:selection];
                        resultText = [@"Won, iOS selected " stringByAppendingString:
computerSelection];
            }

        [resultLabel setText:resultText];

        [buttonView removeFromSuperview];
        [self.view addSubview:resultView];

}
-(void)continueGame:(id)sender{

        [resultView removeFromSuperview];
        [self.view addSubview:buttonView];
}
```

The bold sections of Listing 2–4 indicate where we have made the required changes. We will use this UIViewController version of Rock, Paper, Scissors toward the end of this chapter, after we have set up the rest of our UI.

Customizing Behavior Based on Device Type

As mentioned, the project we are using is an example of a universal application, configured to run on both the iPhone and the iPad. Because it is likely that an application will want to do different things when running on different devices we will create subclasses of our ViewController specific to each device type.

To create these subclasses, from the file menu, select new file. You will be presented with a dialog like the one shown in Figure 2–9.

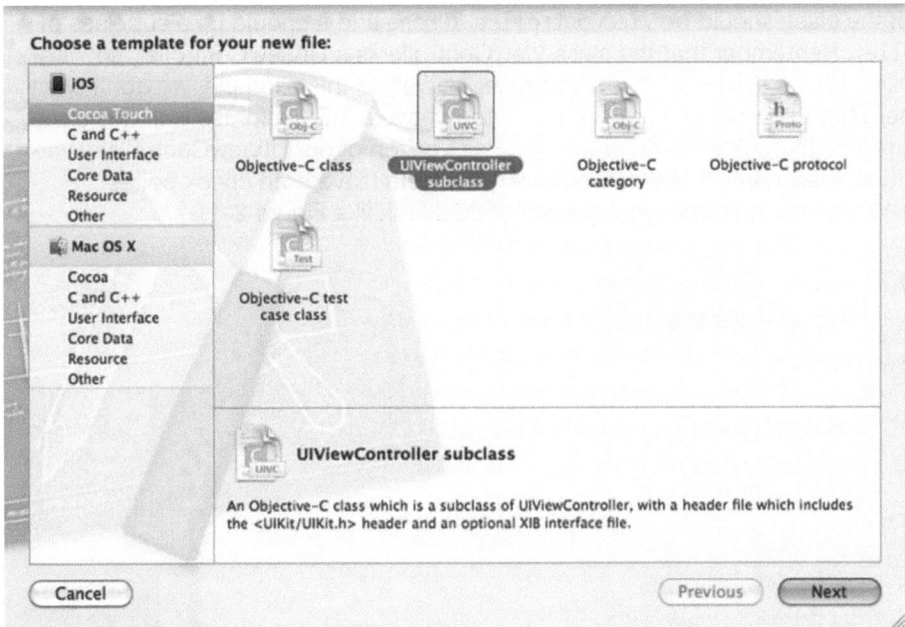

Figure 2–9. *New file dialog*

From the Cocoa Touch section we want to select UIViewController subclass and hit next. This will allow use to name our new class and pick as specific subclass, as shown in Figure 2–10.

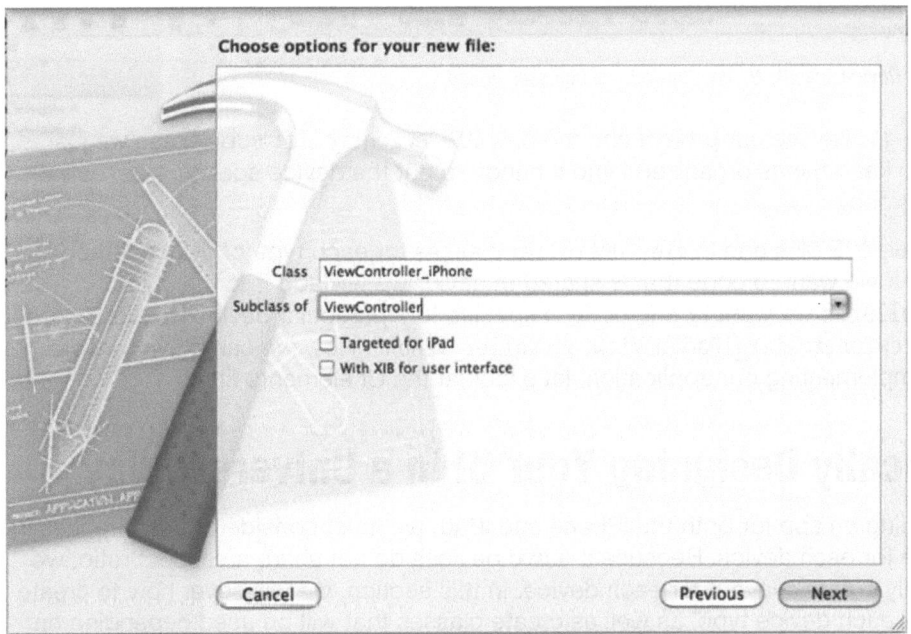

Figure 2–10. *Details for creating a new UIViewController class*

The name of the class should be ViewController_iPhone and it should be a subclass of ViewController. Remember that the class ViewController is a UIViewController, so ViewController_iPhone will be a UIViewController as well. In this example, we don't want to have either check box checked, since we already have a XIB file to use with this class. We want to repeat the process and create an iPad version of our UIViewController. When you create that class, name it UIViewController_iPad and leave both check boxes unchecked. When you are done, your project should look like Figure 2–11.

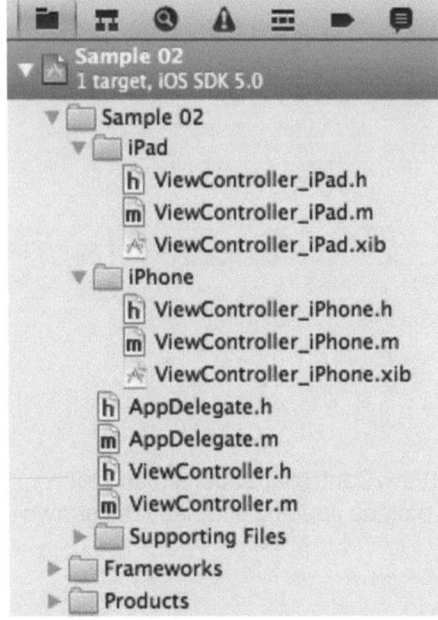

Figure 2–11. *Device specific UIViewController subclasses added*

In Figure 2–11, we see our project and the new UIViewController subclasses we just created. To keep things organized I find it handy to put the device specific classes in their own group.

Now we have XIB files and UIViewController classes for each type of device, iPhones and iPads. If we want to code that is shared behavior we will add it to the class ViewController. If we want to add code is specific to a particular device we would add it to either ViewController_iPad or ViewController_iPhone. Now we can move forward and start implementing our application, let's look at the UI elements first.

Graphically Designing Your UI in a Universal Way

When building an app for both the iPhone and iPad, we must consider the differences in screen size for each device. Because the two devices do not share an aspect ratio, we should really create a layout for each device. In this section, we will cover how to create a layout for each device type, as well as create classes that will be used depending on which device the application is running on.

Xcode provides a convenient and powerful tool for laying out the graphical elements of an application. Historically, that was done with the application Interface Builder, which was a stand-alone application. In recent versions of Xcode, the functionality of Interface Builder has been rolled into Xcode to provide a seamless development environment. Although Interface Builder is no longer its own application, I will continue to use the term to refer to the UI layout tools within Xcode. This will help us distinguish the code editing portions of Xcode from the WYSIWYG elements of Xcode.

At its heart, Interface Builder is a tool for creating a collection of objects and a set of connections between those objects. These objects are UI components and objects that specify behavior or data. This collection of objects is saved in something called an XIB file. XIB files are read at runtime in order to instantiate the objects defined within. The objects that are produced are "wired up" and ready to be used by the application. For example, in Interface Builder, you can add a button to the scene and specify that it should call a task on a specific object when clicked. This is handy, because you don't have to write the code that links the button with the handler object; everything is set up just as you defined it in the XIB file.

> **NOTE:** When searching for help on the Internet regarding Interface Builder, keep in mind that XIB files used to be called NIB files. Much of the information available uses this old term, but is still valid. The same is true of Interface Builder; articles on the old version still give valuable insight into the new version.

A First Look at Interface Builder

Under the iPhone group in the Project Explorer is a file called `ViewController_iPhone.xib`. This file contains the starting visual components used in this project when the application is running on an iPhone. Similarly, there is a file called `ViewController_iPad.xib` found under the iPad group, which is used when the application is running on an iPad. Let's focus on the iPhone version for now to see what this file is and how it works. Click `ViewController_iPhone.xib` and you should see something like Figure 2–12.

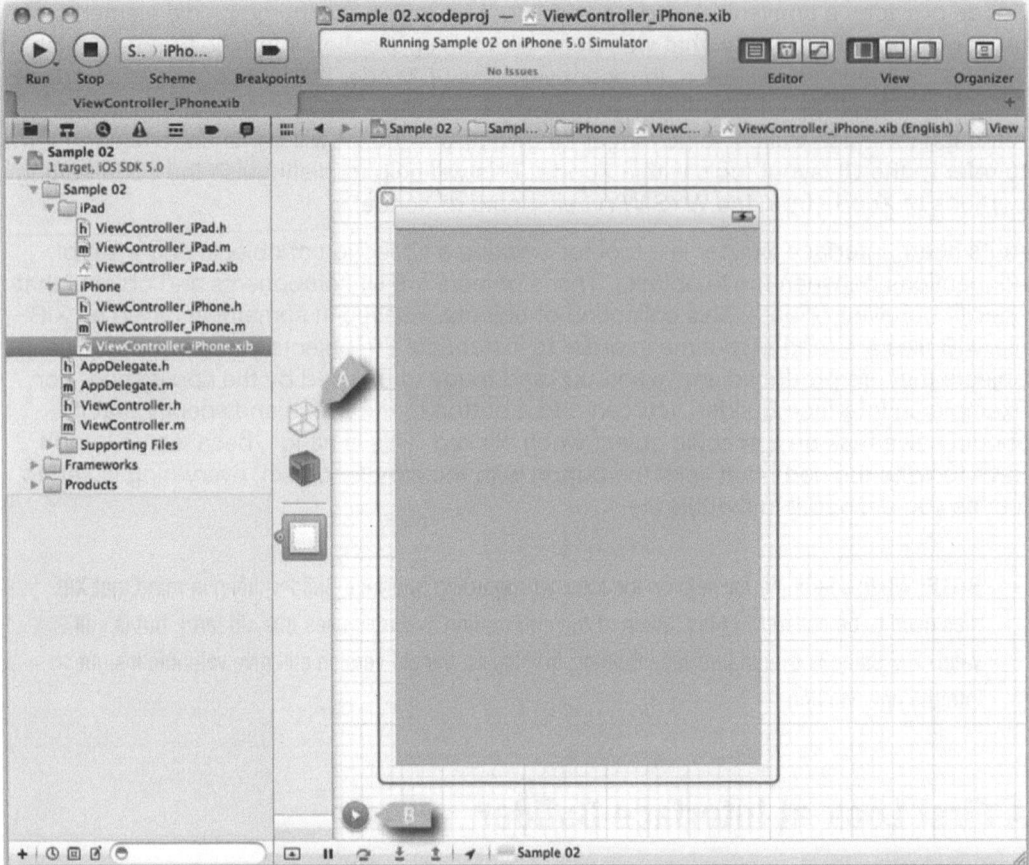

Figure 2–12. *Interface Builder ready to edit ViewController_iPhone.xib*

In Figure 2–12, just to the right of the Project Explorer, is a view with four icons (A). These icons represent the root items in the XIB file. For more detail about the objects, click the little arrow at the bottom of the view (B). This will change the display so it matches the left most view in Figure 2–13. As the label indicates, the items under Object are the objects defined within this XIB file. The items under Placeholders describe relationships to objects not defined in this file. In practice, the item File's Owner is a reference to the object that loads this XIB file; this is most often a subclass of UIViewController, but more on that later.

The other important feature of Figure 2–13 is the area on the right. It shows a graphical representation of the UI components. On the header of Xcode, there are three buttons above the label View (A). When working with Interface Builder, I find it helpful to deselect the left button and select the right button. This will expose a view that shows the properties of the selected items under Objects. When you are done configuring Xcode's UI, you should see something like Figure 2–13.

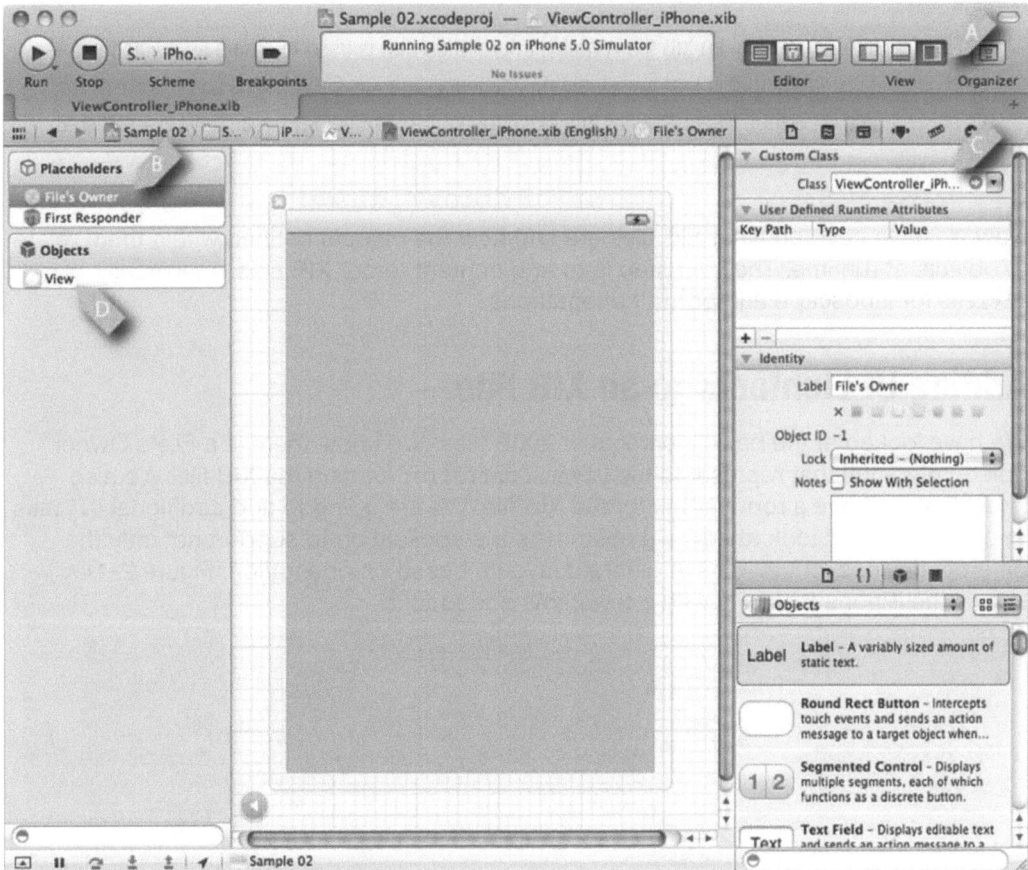

Figure 2–13. *Xcode configured for UI work*

Under the Objects section of Figure 2–10, there is a single item named View (D). This View is a UIView and will be the root view when the application is running. In the center of Figure 2–10, you can see how this view will look.

Under the section Placeholders we see two items. We don't have to concern ourselves with the First Responder at this point. We do however want to select File's Owner (B) and make a modification to it. The File's Owner item is a reference to the object that loads this XIB file. In our case, it will be an instance of ViewController_iPhone, as shown in Listing 2–2. To let Interface Builder know that we intend to use a specific subclass of UIViewController, we enter the class name we want to use on the right side at item C. In this way, Interface Builder will expose the IBOutlets of the class ViewController_iPhone. We will explore more about IBOutlets shortly.

> **TIP:** Technically speaking, an XIB file contains information that is used to create an instance of a particular class. In practice, I find myself simply thinking, "the XIB file contains an object," which is not technically correct. Regardless of this inaccuracy, the terminology police have not arrested me—yet.

Now we have covered what XIB files are and how the objects defined within them relate to objects at runtime. The next step is to add content to our XIB files, specifically UIViews for landscape and portrait orientations.

Adding UI Elements to an XIB File

We have looked at the basic elements in a XIB files. We know there is a File's Owner object reference that represents the UIViewController loading the XIB file. We also know that we have a root UIView for the XIB file. We are going to add additional UIViews to our XIB file and look at how we customize the application to support not only the different devices, but also have different layouts based on orientation. Figure 2–14 shows the iPhone's XIB file with a few items added to it.

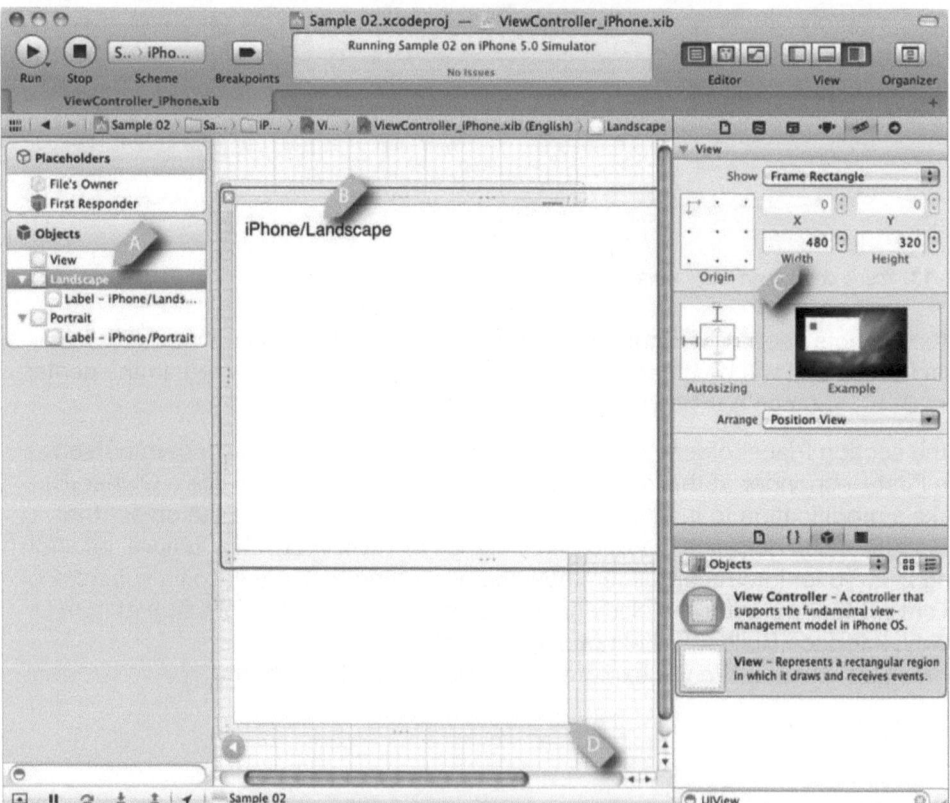

Figure 2–14. *iPhone's XIB file partially configured*

In Figure 2–14, we see the XIB file with an additional UIView objects added. To add new component you simply drag it from the lower right box to the either Objects list or directly onto the scene in the middle. In this case we added two UIView objects named Landscape and Portrait as siblings to the UIView named View (A). Collectively we will refer to these two UIView objects as the orientation views. On each of these new UIView objects we added a UILabel by dragging it from the library at the lower right into the visual representation on the middle of the screen (B). The text for each label will identify which view we are currently displaying. To make sure our layout is correct the size of the Landscape and Portrait UIView objects is set in the properties panel to the right (C). In the center of the screen, you can see the Landscape UIView displayed over the Portrait UIView (D).

The next step is to add a new UIView to each orientation view. This new UIView will describe where our Rock, Paper, Scissors game will be displayed. Figure 2–15, shows this new UIView added to the Portrait orientation view.

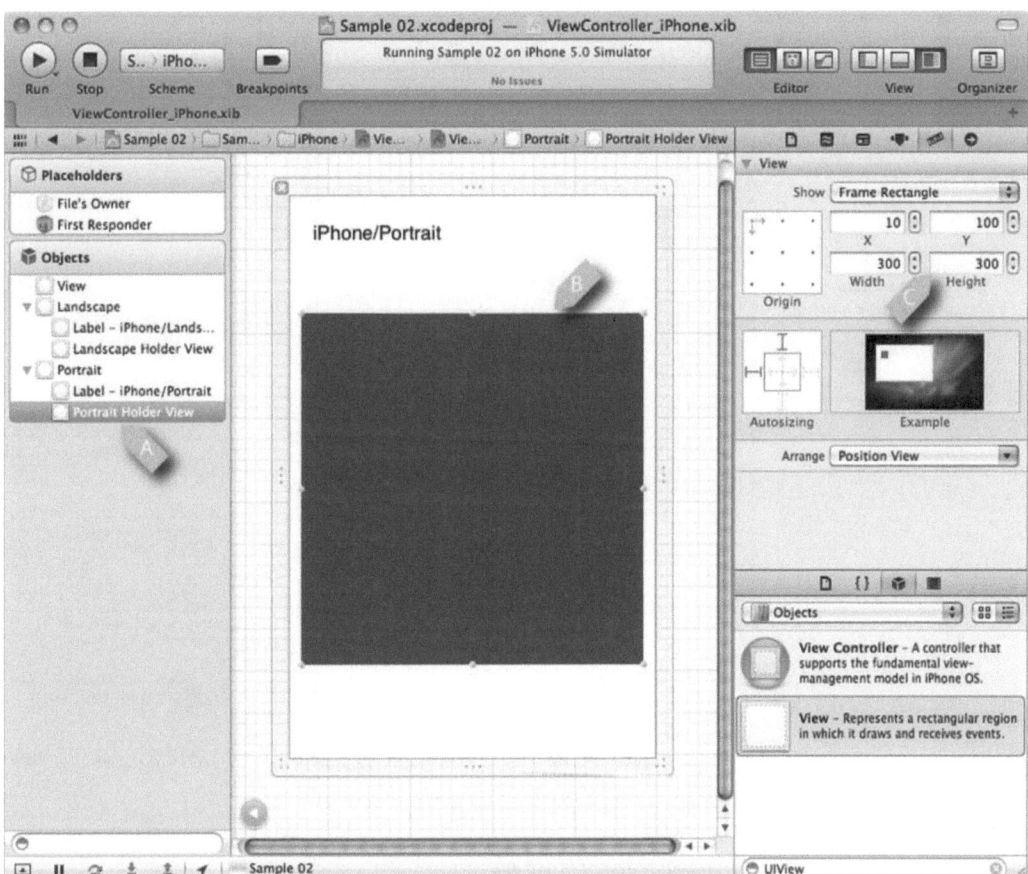

Figure 2–15. *Holder views added to the iPhone's XIB file*

In Figure 2–15, we see that we have added two new UIView objects to the XIB file. The first one is called Portrait Holder View (A) and the second is called Landscape Holder View. In the center of the screen (B) we see that Portrait Holder View is 300x300 (C) points in size and a darker color (red). Landscape Holder View is a sub view of Landscape (not shown).

Add a UIViewController to a XIB

There is one last item we need to add to the XIB file before we are done adding items. We need to add a UIViewController that is responsible for the sub-views that will be displayed in the holder views we just added. The UIViewController we are going to add will be of the class RockPaperScissorsController as defined earlier in this chapter. Figure 2–16 shows the XIB file with a new UIViewController added.

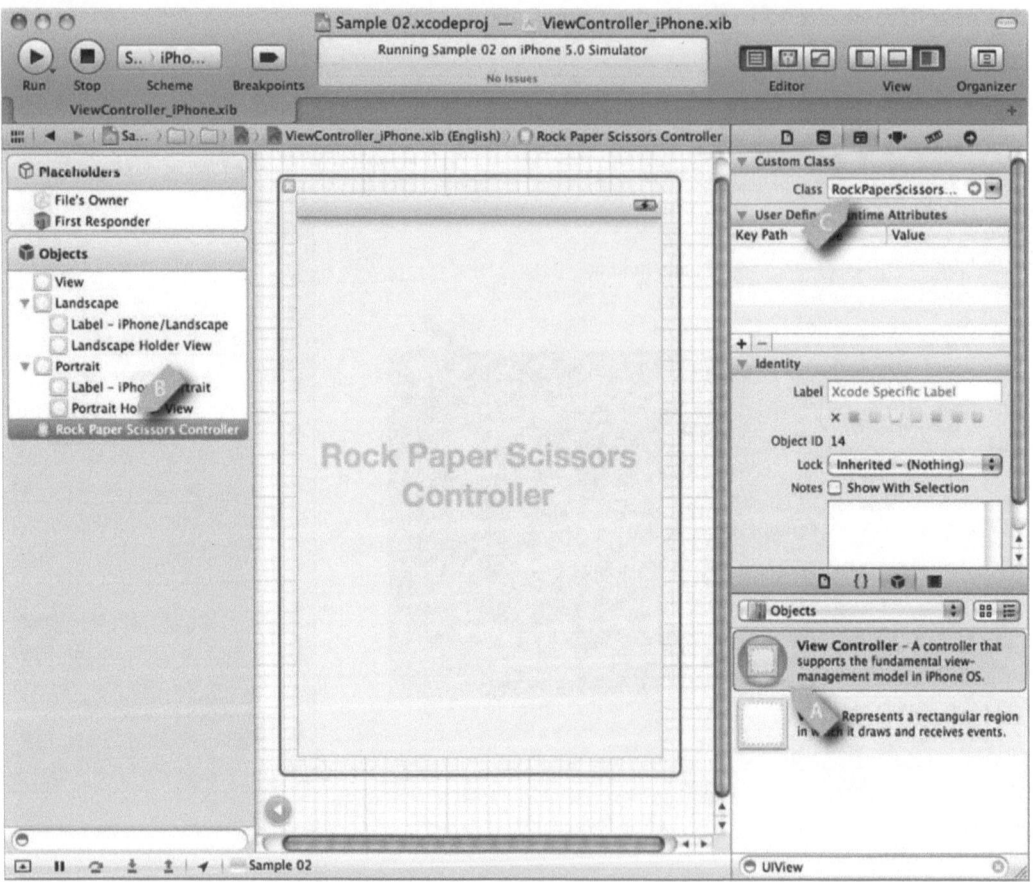

Figure 2–16. *RockPaperScissorsController added to XIB file*

In Figure 2–16, we see that a View Controller was dragged from the library (A) to the Objects section (B) and was set to be the class RockPaperScissorsController (C). In this way, when this XIB file is instantiated, an instance of RockPaperScissorsController will be created as well. This raises the question of how do we access the items within a XIB file programmatically? The answer lies in creating IBOutlets, as the next section discusses.

Creating New IBOutlets from Interface Builder

An IBOutlet is a connection between an item defined in a XIB file and a variable declared in a class. To specify a field as an IBOutlet, you simply put the keyword before the declaration of a field in a classes header file. Another way to create an IBOutlet is to create it through Interface Builder. Figure 2–17, shows creating an IBOutlet in Interface Builder.

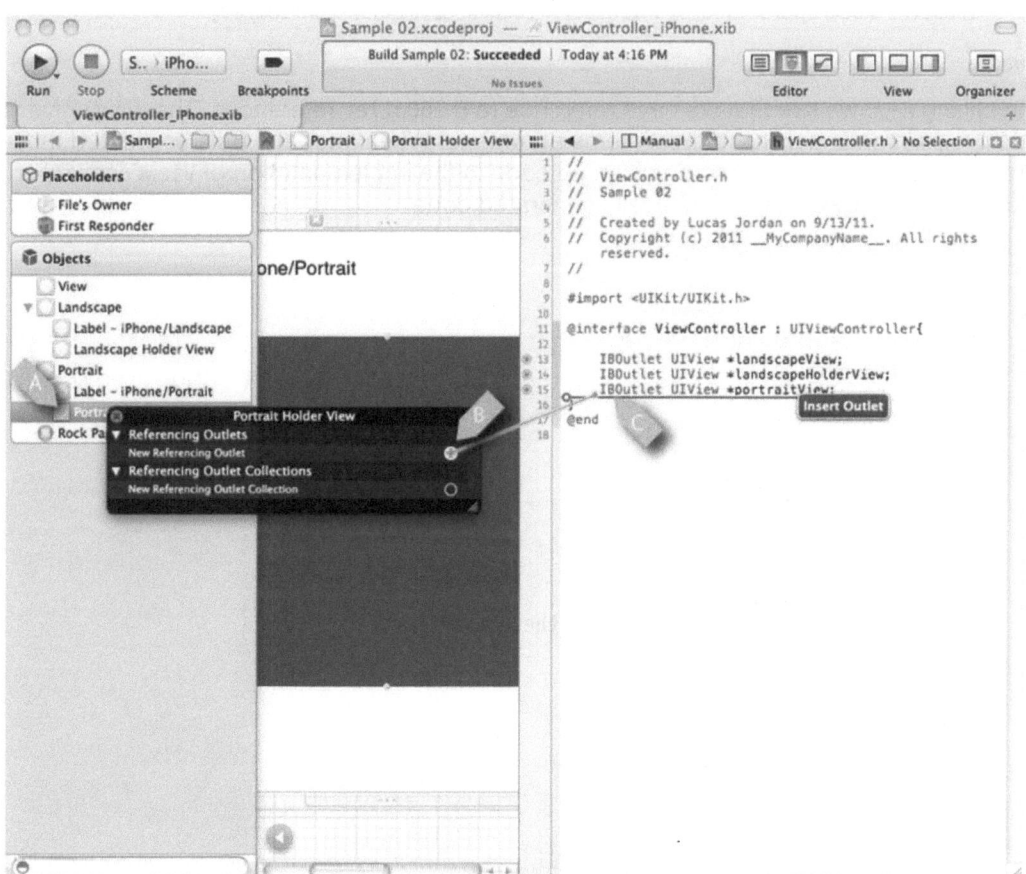

Figure 2–17. *Creating an IBOutlet in Interface Builder*

In Figure 2–17, we see the steps to create an IBOutlet in Interface Builder. This is done by right clicking on the object you want to create an IBOutlet for (A), in this case the Portrait Holder view. From the dialog that pops-up, drag from the little circle to the right of New Referencing Outlet (B) to the place in the code where you want the IBOutlet created (C). As can be seen in Figure 2–17, a number of IBOutlet references are already created. Note that the header file where these IBOutlet references are created is ViewController.h. We use this file ViewController_iPhone and ViewController_iPad will inherit them. Lets take a look at the complete version of ViewControler.h in Listing 2–5.

Listing 2–5. *ViewController.h*

```objc
#import <UIKit/UIKit.h>
#import "RockPaperScissorsController.h"

@interface ViewController : UIViewController{
    IBOutlet UIView *landscapeView;
    IBOutlet UIView *landscapeHolderView;
    IBOutlet UIView *portraitView;
    IBOutlet UIView *portraitHolderView;
    IBOutlet RockPaperScissorsController *rockPaperScissorsController;
}
@end
```

In Listing 2–5, we see that we have reference to 5 IBOutlet references. These will give us programmatic access to these items at runtime. When Interface Builder created these references it also created some clean up code in the implementation of ViewController. Listing 2–6, shows this automatically generated code.

Listing 2–6. *ViewController.m (partial)*

```objc
- (void)viewDidUnload
{
    [landscapeView release];
    landscapeView = nil;
    [landscapeHolderView release];
    landscapeHolderView = nil;
    [portraitView release];
    portraitView = nil;
    [portraitHolderView release];
    portraitHolderView = nil;
    [rockPaperScissorsController release];
    rockPaperScissorsController = nil;
    [super viewDidUnload];
    // Release any retained subviews of the main view.
    // e.g. self.myOutlet = nil;
}
///...
- (void)dealloc {
    [landscapeView release];
    [landscapeHolderView release];
    [portraitView release];
    [portraitHolderView release];
    [rockPaperScissorsController release];
    [super dealloc];
}
```

In Listing 2–6, we see code generated by Interface Builder for each IBOutlet created. In the task viewDidUnload we see that each view is released and set to nil. Similarly, in the task dealloc we see that each view is released. In the next section we are going to review how we respond to orientation change, this same code will be responsible for getting the different views have been working with onto the screen.

Responding to Changes in Orientation

The last thing we have to do is add some logic that changes which UIView is displayed based on the orientation of the device. The class UIViewController specifies a task that is called when the device rotates. This tasks is called shouldAutorotateToInterfaceOrientation: and it is passed a constant indicating the new orientation of the device. By adding logic to this task, we can implement the type of behavior we want.

When a new UIViewController is created, an implementation of this task is included in the implementation file. The work we did creating the subclasses of the ViewController class and wiring up the XIB files allows us to use a single implementation of the shouldAutorotateToInterfaceOrientation: task. So open ViewController_iPad.m and ViewController_iPhone.m and delete the shouldAutorotateToInterfaceOrientation: task implementation. This will insure that, on both devices, we will be using the implementation from the class ViewController. Listing 2–7 shows the implementation that gives us the behavior we want.

Listing 2–7. *GameController.m (shouldAutorotateToInterfaceOrientation:)*

```
-
(BOOL)shouldAutorotateToInterfaceOrientation:(UIInterfaceOrientation)interfaceOrientatio
n
{
    if (interfaceOrientation == UIInterfaceOrientationLandscapeLeft ||
interfaceOrientation == UIInterfaceOrientationLandscapeRight){
        [portraitView removeFromSuperview];
        [self.view addSubview:landscapeView];

        [rockPaperScissorsController setup:landscapeHolderView.frame.size];
        [landscapeHolderView addSubview:rockPaperScissorsController.view];

    } else {
        [landscapeView removeFromSuperview];
        [self.view addSubview:portraitView];

        [rockPaperScissorsController setup:portraitHolderView.frame.size];
        [portraitHolderView addSubview:rockPaperScissorsController.view];
    }

    // Return YES for supported orientations
    return YES;
}
```

In Listing 2–7, we see that this task has passed the variable interfaceOrientation, indicating the new orientation of the device. At the very end, we return Yes, because all orientations are supported. If you were creating a game that does not support all orientations, you would return No for the undesired orientations. Before we return a value, we test interfaceOrientation to see if we are in landscape or portrait orientation. Because the device really has four orientations, we have to compare interfaceOrientation to both landscape right and landscape left. The right/left indicator is describing the position of the Home button.

If the device is now in landscape orientation, we remove the portraitView from its super view — meaning, we take it off the screen. We then add the view landscapeView to the root most view. We call setup on rockpaperScissorsController to give it a chance to lay out its subviews, if it has not already. Lastly, we add rockPaperScissorsController's view property to landscapeHolder as a subview.

This task is called when the application starts, even if the device is rotated. This lets us put all of our orientation logic in one place, because we don't have to deal with the special case of initialization. However, depending on whether this is the first time shouldAutorotateToInterfaceOrientation: is called, the state of these objects will be different. For example, the first time this task is called, neither the portraitView nor the landscapeView are attached to the scene, so the call to removeFromSubview will be a non-operation, without error. When this task is called on subsequent rotations, rockPaperScissorsController's view will be attached to the scene; however, when a view is added as a subview, it is automatically removed from its parent view, if it has one.

The fact that removeFromSuperview and addSubview: have these reasonable default behaviors makes life pretty easy when it comes to manipulating views. It cuts down on error-prone clean-up code.

If you run the project, you will see the app running on both the iPad and the iPhone, changing its layout based on orientation.

Summary

In this chapter, we looked at how to create an Xcode project suitable for developing a universal application. We explored the initialization process of an iOS application, and saw where to start modifying the project to create the application we want. We looked at how XIB files and classes interact, and at how iOS exploits MVC with the UIViewController class. We created an application that seamlessly supports running on an iPhone or an iPad in all orientations, setting the stage for game development.

Chapter 3

Explore the Game Application Life Cycle

There is more to a game than just the fun parts. Almost all of the games on the market, and definitely the big titles, involve multiple views and a reasonably complex life cycle. This chapter explores how to turn a simple game into a complete application. We will look at how a typical game organizes the different views that comprise it, including a Loading view, a Welcome view, a High Score view, and of course the game itself.

Beyond user navigation, an application also has a life cycle. On iOS, that includes responding to the application being terminated, recovering from being in the background, or starting up for the first time. When an application moves through its life cycle, it will want to preserve elements of the user data, like high scores and the state of the game. Paying attention to these little details will make your game look polished instead of amateurish and annoying.

While the game presented here is not very exiting and the art is admittedly amateurish, you will be able to take the basic outline presented here and drop your own game into it. You should update your art though!

Understanding the Views in a Game

The game presented in this chapter is called Coin Sorter. It is a simple puzzle game in which the player sorts coins into rows and columns. The player has ten turns to make as many matching rows or columns as possible. The fun part of this game is just about as simple as an iOS game can get. The playable part of this game is comprised of just two classes, CoinsController and CoinsGame. As the names of these two classes suggest, CoinsController is responsible for rendering the game to the screen and interpreting player actions. CoinsGame is responsible for storing the state of the game and providing a few utility methods for manipulating that state. The details of how these two classes work Is explored fully in Chapter 4. In this chapter we are going to look at all of the surrounding code that supports these two classes to create a complete application.

Most games are comprised of several views—not just the view where the main action of the game takes place. These views often include a Welcome view, a High Score view, and maybe a few others. The sample code that accompanies this chapter is designed to outline how these supporting views are managed. Figure 3–1 is a flowchart of the views used in this game.

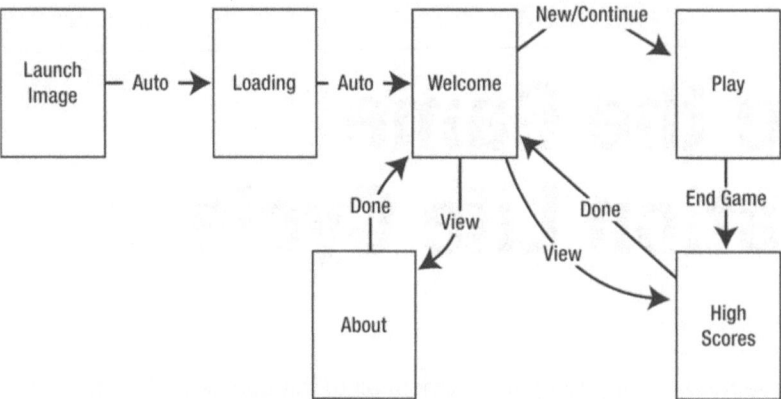

Figure 3–1. *Coin Sorter application flow*

In Figure 3–1, each rectangle is a view in the game. The arrows indicate the transitions between these views. Starting on the left, there is a box with the title Launch Image. This view is different from the others, as it is just a PNG file that is displayed before control of the application is passed to the app delegate. See Appendix A for the details of configuring this image. When the application has fully loaded, we display the Loading view. This view stays on the screen for a few seconds while the application pre-loads some images. Once the images are loaded, we show the user the Welcome view.

The Welcome view in Figure 3–1 offers players a number of options. They can continue an old game, start a new game, view the high scores, or view information about the game. If the user plays the game, they are directed to the Play view, where the main action of this application takes place. When they are done with a game, the user is shown the High Scores view. When the user is done reviewing the high scores, they are sent back to the Welcome view, enabling them to again select from the available options.

The application flow presented in Figure 3–1 is by no means the only option, but it presents a common pattern. Extending or modifying this example should be simple once you understand what each view does and how the state of the application is managed.

> **NOTE:** Although laying out this type of workflow with the Storyboard feature of Xcode might be appealing, supporting multiple orientations and multiple devices makes using Storyboard impractical. Ideally, Storyboard will improve in future versions of Xcode and become the only tool used to manage application life cycles.

Exploring the Role Each View Plays

Each view in the application serves a different purpose. By understanding what each view does, you will understand why this example is organized the way that it is. Let's start by looking at Figure 3–2, showing the launch image and the Loading view.

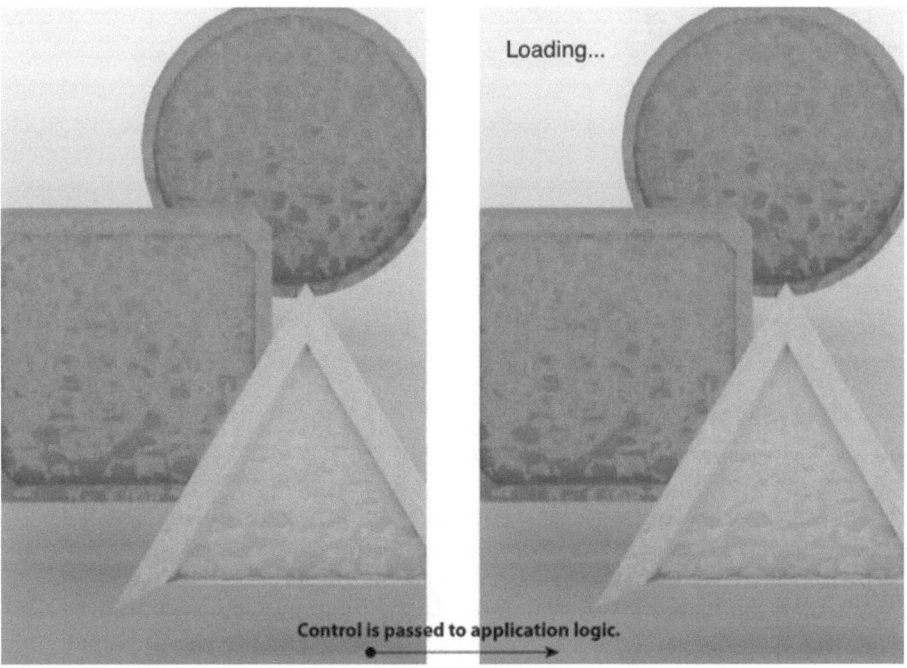

Figure 3–2. *Launch image and Loading view*

In Figure 3–2, we see the launch image on the left, which appears when control of the initialization process is passed to the app delegate. The application shows the Loading view shown on the right of Figure 3–2. In this case, the Loading view has a UIImageView with the same image used as the launch image. This is not required, but it is a nice way to provide a seamless visual experience for the user. In a more complex application, we might want to display a progress bar as images are loading. For an application this simple, we don't really need a loading screen, because the number and size of the images used is small. In our case, we simply add a UILabel with the text "Loading…" to the Loading view. Once the initial resources of an application are loaded, the Welcome view is presented, as shown in Figure 3–3.

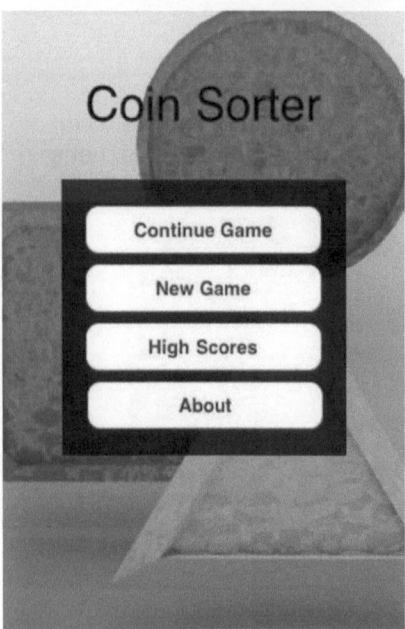

Figure 3–3. *Welcome view showing the Continue Game button*

The Welcome view shown in Figure 3–3 presents four buttons to the player. The first button allows the player to continue the previous game. The New Game button starts a new game . The buttons High Scores and About allow the player to review past high scores and view the About screen. The Welcome view is the home screen for the player. Looking at Figure 3–1, we see all future actions will inevitably lead the player back to this view. If the player chooses to continue an earlier game or start a new game, the Play view is presented, as shown in Figure 3–4.

Figure 3–4 shows the Play view, where the game play actually happens. At the top of the screen, the number of remaining turns is displayed, as well as the current score. The square region with the geometric shapes is where the game takes place. Each triangle, square, and circle represents a different type of coin. The object of the game is to create columns and rows of like coins. The only way to rearrange the coins is by selecting two to trade place. When a row or column of like coins is created, the player's score is incremented and new coins appear to replace those in the matching column or row. The player can only swap coins ten times before the game is over, but with careful planning, a player can create multiple matches per swap. When the game is over, the High Score view is shown, highlighting any new high score the player may have achieved. Figure 3–5 shows the High Score view.

Figure 3–4. *Play View—a game in progress*

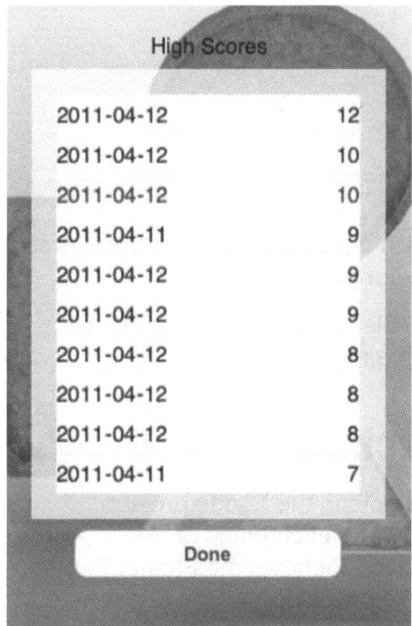

Figure 3–5. *The High Score view*

In Figure 3–5, we see the High Score view, where the third score from the top is the score of the last game played. The high scores presented in this view are saved between application sessions. We will explore how the high score data is save later in this chapter. When the user is done inspecting their past and current success, they can select the Done button to be directed back to the Welcome view. From the Welcome view, players can click the About button and view some information about the game, as shown in Figure 3–6.

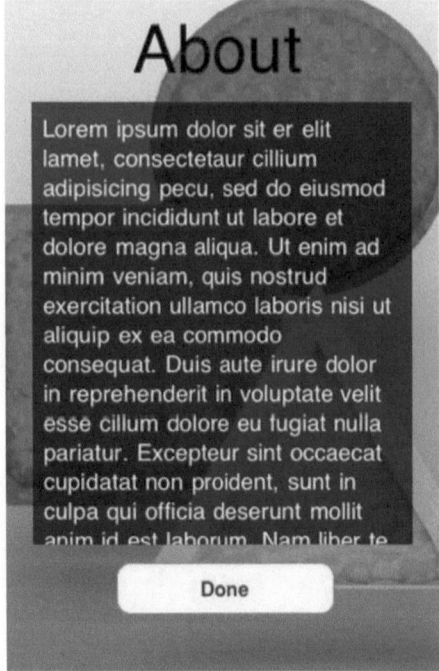

Figure 3–6. *The Welcome view*

The Welcome view presented in Figure 3–6 is very much a placeholder. You would want to replace the "Lorem ipsem..." with a description of your game.

In the Chapter 2, we explored how to configure an XIB file containing the various UI elements in the game. With this game, we follow a very similar pattern.

Understanding the Project's Structure

Even a simple game can have a surprising number artifacts in the Xcode project. Exploring what these items are helps you understand where each piece of the application is defined and how each fits together. Figure 3–7 shows the Project Explorer in Xcode, which shows a number of the files used in this project.

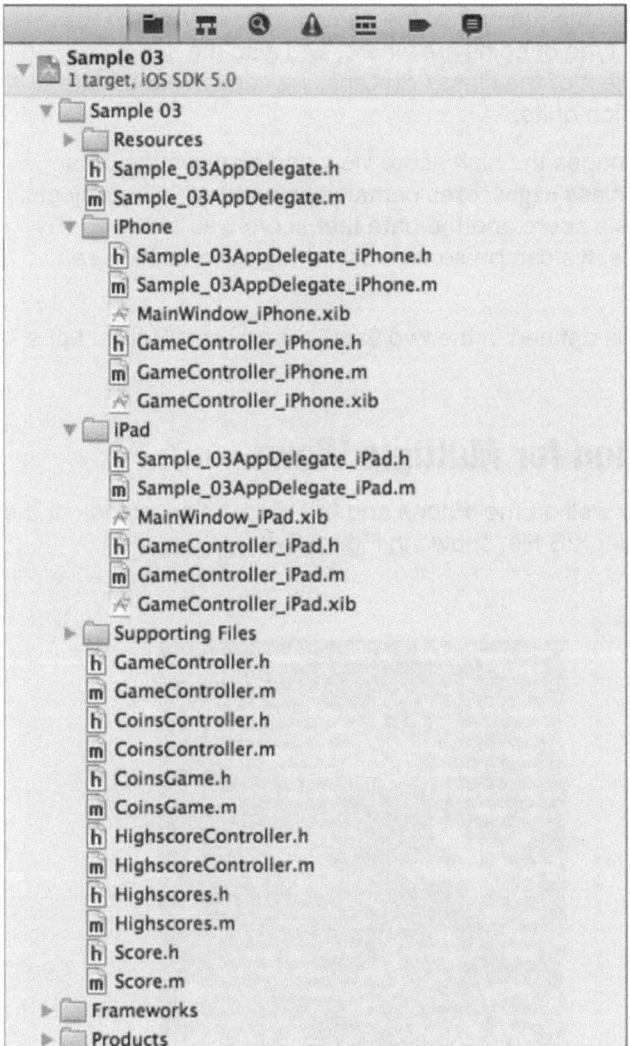

Figure 3–7. *Project Explorer for the Project Sample 03*

The project shown in Figure 3–7 follows the same pattern as the project from Chapter 2. There are the shared classes under the group Sample 03, while device-specific classes are stored under the groups iPhone and iPad. The class Sample_03AppDelegate contains the initialization code for the application and code for saving the user state when a player exits the application. The XIB files are also wired up very much like the XIB files from Chapter 2: the XIB files starting with MainWindow each contain a GameController of the appropriate subclass. The class GameController contains all of the logic for managing the state of the application, while the XIB files for each subclass contain the UI elements suitable for each device.

The class `CoinsController` describes the fun part of the game and the class `CoinGame` is a model class that describes which type of coins are where, and also the score and remaining turns. By archiving an object of the class `CoinGame`, we can save the state of the current game when the application quits.

The class `HighscoreController` manages the high score view and displays the data found in the class `Highscores`. The class `Highscores` contains an array of `Score` objects, where each `Score` object represents a score and the date that score was achieved. By serializing an instance of `HighScores`, we can preserve a user's high scores between play sessions.

The meat of the view configuration is defined in the two `GameController` XIB files. Let's take a look at those.

Configuring an Application for Multiple Views

While this application works equally well on the iPhone and the iPad, let's just look at the iPhone version of the GameController XIB file, shown in Figure 3–8.

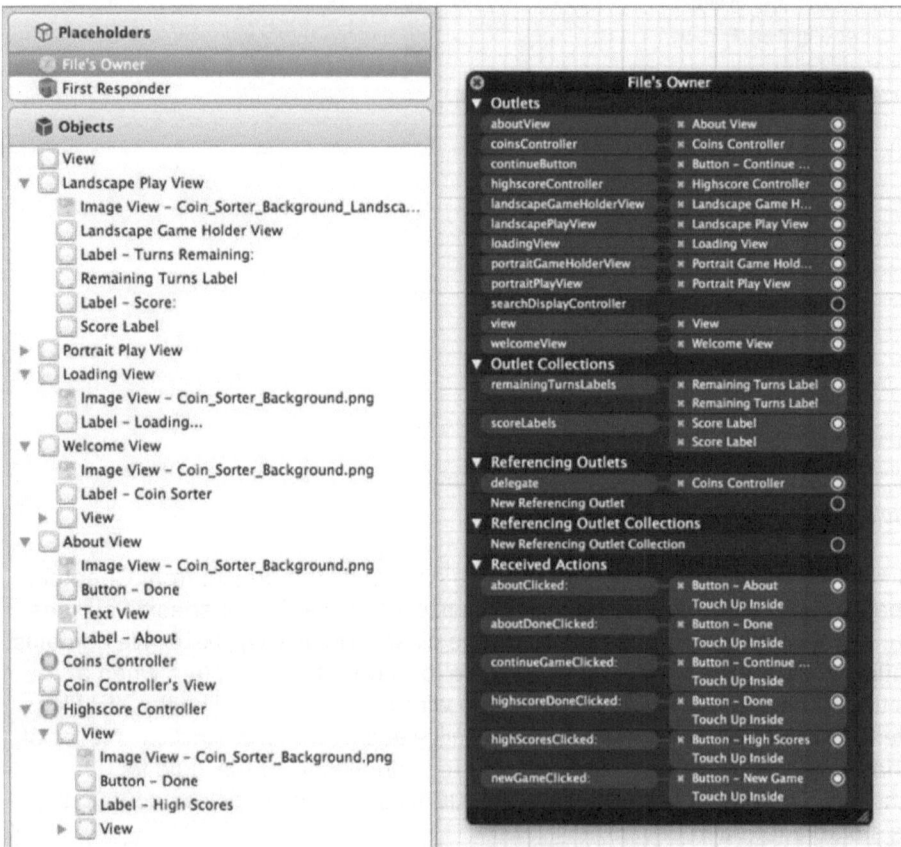

Figure 3–8. *GameController_iPhone.xib*

In Figure 3–8, we see the details of the file GameController_iPhone.xib on the left. On the right, we see the IBOutlets and IBActions defined in the file GameController.h.

Reviewing GameController_iPhone.xib

Under the Objects section on the left, we see a number of UIViews. The Landscape Play view and the Portrait Play view are shown when the game is being played, which is the only time this application supports both orientations. For simplicity, the rest of the views are for portrait orientation only.

When the application initializes, we know that the first view the user sees is the Loading view. After that, the user is directed to the Welcome view. In Figure 3–8, we can see that both of these views are defined in the XIB file. We also see that the About view is defined in the XIB file. For the actual game and the High Score view, we see corresponding UIViewControllers in the XIB. These two views are handled with a UIViewController because the each contributes a reasonable amount of functionality, so it makes sense to define their behaviors in there own UIViewController classes. As a side effect of this design decision, we can reuse high score UIViewController in other applications.

Reviewing GameController.h

On the right of Figure 3–8, we see the connections defined in the class GameController. We can also see that most of the Objects listed are wired up to various connections. Wiring up these components in this way allows us to easily reference these different components in code. Listing 3–1 shows the definition of these connections in GameController.h

Listing 3–1. *GameController.h*

```
#import <UIKit/UIKit.h>
#import "CoinsController.h"
#import "CoinsGame.h"
#import "HighscoreController.h"

@interface GameController : UIViewController <CoinsControllerDelegate>{

    IBOutlet UIView *landscapePlayView;
    IBOutlet UIView *landscapeGameHolderView;
    IBOutlet UIView *portraitPlayView;
    IBOutlet UIView *portraitGameHolderView;
    IBOutlet UIView *loadingView;
    IBOutlet UIView *welcomeView;
    IBOutlet UIButton *continueButton;
    IBOutlet UIView *aboutView;

    IBOutlet CoinsController *coinsController;
    IBOutletCollection(UILabel) NSArray *remainingTurnsLabels;
    IBOutletCollection(UILabel) NSArray *scoreLabels;
    IBOutlet HighscoreController *highscoreController;
```

```
        CoinsGame* previousGame;
        BOOL isPlayingGame;
}
-(void)setPreviousGame:(CoinsGame*)aCoinsGame;
-(CoinsGame*)currentGame;

- (IBAction)continueGameClicked:(id)sender;
- (IBAction)newGameClicked:(id)sender;
- (IBAction)highScoresClicked:(id)sender;
- (IBAction)aboutClicked:(id)sender;
- (IBAction)aboutDoneClicked:(id)sender;
- (IBAction)highscoreDoneClicked:(id)sender;

-(void)loadImages;
-(void)showPlayView: (UIInterfaceOrientation)interfaceOrientation;

@end
```

In Listing 3–1, we import three other classes defined in this project, including the default import statement for UIKit. The rest of the IBOutlet statements are basic declarations, allowing a number of different views to be accessed. The IBOutletCollection definitions represent a new type of connection for us. An IBOutletCollection acts very much like an IBOutlet, except it allows you to connect multiple items to a single NSArray. In our case, we have two labels for displaying the score: landscape and portrait. Instead of having an IBOutlet (and variable) for each of these, we can simply use an IBOutletCollection, so the NSArray scoreLabels will contain both of these labels at runtime. When it is time to update the UI with a new score, we simply update the text of all of the UILabels in the array scoreLabels. We follow the same pattern for the labels that display the remaining turns, where each is stored in the NSArray remainingTurnsLabels.

Bringing the Views onto the Screen

That completes our overview of how the different views are defined. Now let's take a look at how each of these views is brought onto the screen, starting with the Loading view.

We know that the Main view for the class GameController is displayed when the application loads, just like it did in Chapter 2. However, the Main view for GameController is empty—it is effectively the root component for our application. When we change views in the application, we will be adding and removing views from this root view. Listing 3–2 shows the viewDidLoad task from GameController, where we add the first view we want the user to see.

Listing 3–2. *GameController.m (viewDidLoad)*

```
- (void)viewDidLoad
{
    [super viewDidLoad];
    [self.view addSubview:loadingView];

    NSOperationQueue *queue = [NSOperationQueue mainQueue];
    NSInvocationOperation *operation = [[NSInvocationOperation alloc]
initWithTarget:self selector:@selector(loadImages) object:nil];
```

```
    [queue addOperation:operation];
    [operation release];
}
```

The viewDidLoad tasks, as shown in Listing 3–2, is called when an instance of GameController is loaded from an XIB file. Because an instance of GameController is in each MainWindow XIB file, this happens as soon as the application starts. After calling the super implementation of viewDidLoad, which is required, we add the UIView loadingView to the main view of the GameController class. The task addSubview: displays the new sub-view at the location and size as define by the sub-views frame property. Because loadingView is defined in the XIB file, its frame is defined there.

After we get the Loading view on the screen, we want to load any initial resources so we can prevent an untimed pause later in the application life cycle. It is not technically required in this app to load the images asynchronously, but if we wanted to display a progress bar, this would be required. In order to perform a task asynchronously, we get access to the main NSOperationQueue by calling mainQueue. An NSOperationQueue provides a context for our code to perform work on another thread. The class NSInvocationOperation is a way to define a unit of work. In our case, the unit of work is calling the task loadImages on this GameController object. So we initialize an NSInvocationOperation by calling the task initWithTarget:selector:object: and passing self, a selector for the task loadImages, and nil. Lastly, we get the execution started by calling addOperation: and passing the NSInvocationOperation. The loadImage task is shown in Listing 3–3.

Listing 3–3. *GameController.m (loadImage)*

```
-(void)loadImages{
    //sleeping so we can see the loading view. In a production app, don't sleep :)
    [NSThread sleepForTimeInterval:1.0];

    [coinsController loadImages];
    [loadingView removeFromSuperview];
    [self.view addSubview:welcomeView];
}
```

The loadImage task calls sleepForTimeInterval and causes the current thread to sleep for a second. This is done to make sure you can actually see the loading view in the sample code; delete anything like this in production code. After the sleep, call loadImages on coinsController. The details of CoinsController's implementation of loadImages is described in Chapter 4—all you need to know now is that it loads all of the images used to display the coins. When we are done loading, we remove the loadingView and add the welcomeView.

The Welcome view is really where the application starts for the user. From there they can access the interesting parts of the application.

Changing Views in Response to User Actions

Once the Welcome view is displayed, the user is presented with a number of buttons. This view is shown in Figure 3–3. The top button, Continue Game, is only present if the user was partially through a game when they exited the application. Each button causes the user to navigate to a new view. This is done by first establishing a task that will be called when the user touches a button. In Listing 3–1, we see a number of tasks declared in the GameController.h file. The ones starting with IBAction are tasks that are called when a button is touched. Once the tasks are declared in the header file, you can wire up each button to each task. Figure 3–9 shows the first step in wiring up a button.

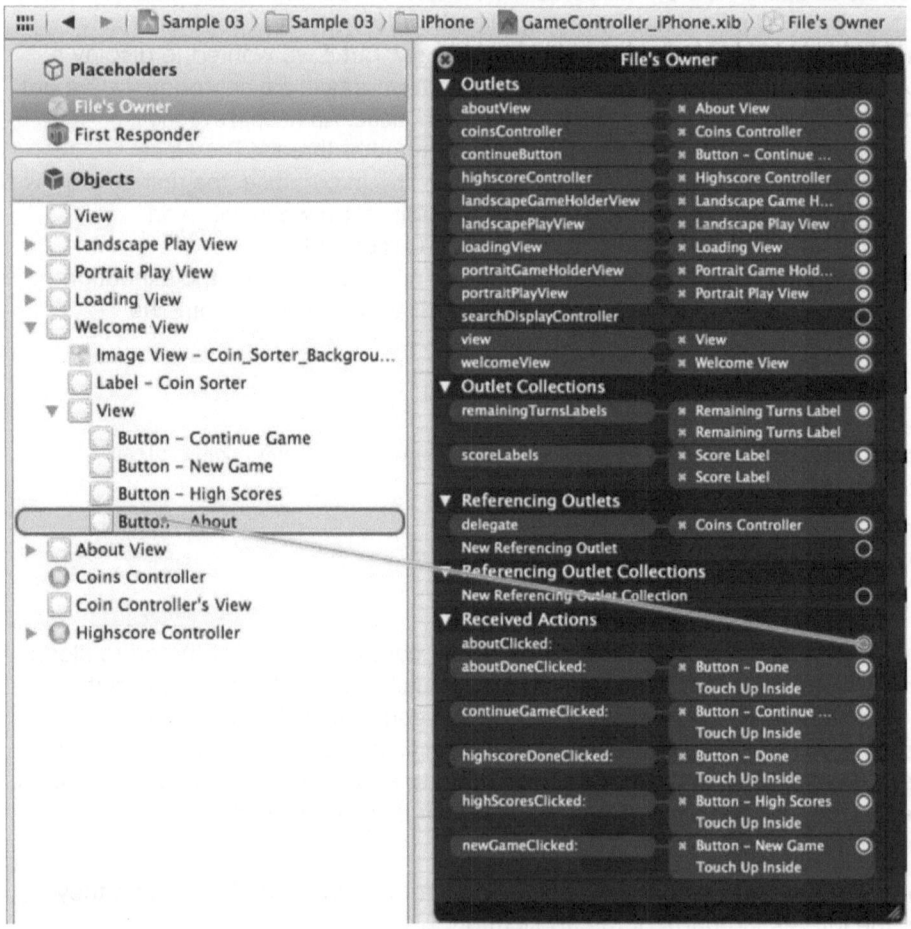

Figure 3–9. *Wiring the aboutClicked: action to the About button*

In Figure 3–9, we see the Connections menu of File Owner exposed by right-clicking on File Owner. By dragging from the circle to the right of aboutClicked: to the correct button on the left, you can select the action that triggers the task aboutClicked:. When you let go of the button, a second pop-up menu is presented, allowing you to select the specific user action. Figure 3–10 shows the selection of the action Touch Up Inside, which is typically the desired action for a button.

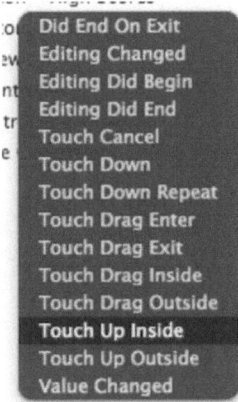

Figure 3–10. *Selecting the action Touch Up Inside*

Once a button's action is wired up in the XIB, you have to provide and implement the task for the button to do anything. In our sample application, we want each button to change the current view. Listing 3–4 shows the implementation of the these action tasks in GameController.m.

Listing 3–4. *GameController.m (IBAction tasks)*

```
- (IBAction)continueGameClicked:(id)sender {
    [UIView beginAnimations:nil context:@"flipTransitionToBack"];
    [UIView setAnimationDuration:1.2];
    [UIView setAnimationTransition:UIViewAnimationTransitionFlipFromRight
forView:self.view cache:YES];

    [welcomeView removeFromSuperview];

    UIInterfaceOrientation interfaceOrientation = [self interfaceOrientation];
    [self showPlayView:interfaceOrientation];

    [coinsController continueGame:previousGame];

    [UIView commitAnimations];
    isPlayingGame = YES;
}

- (IBAction)newGameClicked:(id)sender {
    [UIView beginAnimations:nil context:@"flipTransitionToBack"];
    [UIView setAnimationDuration:1.2];
    [UIView setAnimationTransition:UIViewAnimationTransitionFlipFromRight
forView:self.view cache:YES];

    [welcomeView removeFromSuperview];

    UIInterfaceOrientation interfaceOrientation = [self interfaceOrientation];
    [self showPlayView:interfaceOrientation];

    [coinsController newGame];
```

```
        [UIView commitAnimations];
        isPlayingGame = YES;

}

- (IBAction)highScoresClicked:(id)sender {
        [UIView beginAnimations:nil context:@"flipTransitionToBack"];
        [UIView setAnimationDuration:1.2];
        [UIView setAnimationTransition:UIViewAnimationTransitionFlipFromRight
forView:self.view cache:YES];

        [welcomeView removeFromSuperview];
        [self.view addSubview:highscoreController.view];


        [UIView commitAnimations];
}

- (IBAction)aboutClicked:(id)sender {
        [UIView beginAnimations:nil context:@"flipTransitionToBack"];
        [UIView setAnimationDuration:1.2];
        [UIView setAnimationTransition:UIViewAnimationTransitionFlipFromRight
forView:self.view cache:YES];

        [welcomeView removeFromSuperview];
        [self.view addSubview: aboutView];


        [UIView commitAnimations];
}
```

In Listing 3–4, we see all of the tasks used to control which view we show from the Welcome view. The first three lines of each task sets up the animation that will be used to transition from one view to another. The last line that calls UIView's commitAnimations task indicates that you have specified all of the UI changes you want animated, and that the animation should be shown to the user. The type of animation is controlled by setting the context string on UIView's beginAnimations:context: task. In order actually change which view will be displayed, we simply have welcomeView remove itself from its parent by calling removeFromSuperview. Then we add the desired view to self.view by calling the task addSubview:.

For the High Score view and the About view, we want the user to navigate back to the Welcome view. Listing 3–5 shows these action tasks.

Listing 3–5. *GameController.m (Navigate Back to the Welcome View)*

```
- (IBAction)aboutDoneClicked:(id)sender {
        [UIView beginAnimations:nil context:@"flipTransitionToBack"];
        [UIView setAnimationDuration:1.2];
        [UIView setAnimationTransition:UIViewAnimationTransitionFlipFromLeft
forView:self.view cache:YES];

        [aboutView removeFromSuperview];
        [self.view addSubview: welcomeView];

        [UIView commitAnimations];
```

```
}

- (IBAction)highscoreDoneClicked:(id)sender {
    [UIView beginAnimations:nil context:@"flipTransitionToBack"];
    [UIView setAnimationDuration:1.2];
    [UIView setAnimationTransition:UIViewAnimationTransitionFlipFromLeft
forView:self.view cache:YES];

    [highscoreController.view removeFromSuperview];
    [self.view addSubview: welcomeView];

    [UIView commitAnimations];
}
```

Listing 3–5 shows two tasks that are called when the user touches the Done buttons on the About view and the High Score view. These tasks are very similar to the tasks in Listing 3–4. The only real difference is that we remove the About and High Score views and then add the Welcome view. This is a reversal from the tasks in Listing 3–4.

When the user navigates to the Play view, they stay on that view until their game is done. This application is set up to automatically change views in this case. The next section explains how GameController interacts with the CoinsController, not only to know when to transition the user to the High Score view, but also when to update the remaining turns and score labels.

Using a Delegate to Communicate Application State

A common pattern for one object to communicate with another in iOS development is the delegate pattern. A delegate is simply any object that conforms to some predefined protocol that another object knows about. For example, say we have class A and it has a property called delegate that must conform to the protocol P. We can assign an instance of class B to the delegate property of A, if class B conforms to the protocol P. The delegate pattern is very much like the listener pattern found in other languages, like Java.

Declaring CoinsControllerDelegate

In order for our CoinsController class to communicate its state with the class GameController, we create a protocol that describes this relationship. Listing 3–6 shows the declaration of the protocol CoinsControllerDelegate found in the file CoinsController.h.

Listing 3–6. *CoinsControllerDelegate (from CoinsController.h)*

```
@protocol CoinsControllerDelegate <NSObject>

-(void)gameDidStart:(CoinsController*)aCoinsController with:(CoinsGame*)game;
-(void)scoreIncreases:(CoinsController*)aCoinsController with:(int)newScore;
-(void)turnsRemainingDecreased:(CoinsController*)aCoinsController
with:(int)turnsRemaining;
-(void)gameOver:(CoinsController*)aCoinsController with:(CoinsGame*)game;

@end
```

As can be seen in Listing 3–6, a protocol is a pretty simple thing. The name of the protocol comes after the @protocol. The <NSObject> specifies what types of objects can conform to this protocol. Between the @protocol and the @end, a number of tasks are declared. These are the messages that are going to be sent to the delegate object, so the delegate object should implement each of these tasks. The delegate field is declared in the interface section of the CoinsController like this:

```
IBOutlet id<CoinsControllerDelegate> delegate;
```

We also define a property for this field, with the property declaration:

```
@property (nonatomic, retain) id<CoinsControllerDelegate> delegate;
```

There are two things to notice about the declaration of the delegate property. First, you can assign any object as the delegate. Because it is type id, you will get a compiler warning if the assigned object is known not to conform to the protocol CoinsControllerDelegate. The second thing to notice is that we defined it as an IBOutlet, which allowes us to actually set the relationship between the GameController object and the CoinsController object in Interface Builder. This is not required, but it can be a handy addition if you are planning on writing code consumed by other developers.

Looking at the declaration of the class GameController, we can see that it conforms to the protocol CoinsControllerDelegate, like such:

```
@interface GameController : UIViewController <CoinsControllerDelegate>{
```

Implementing the Defined Tasks

Simply stating that a class conforms to a protocol is just the first step. The conforming class should also implement each task defined in the protocol. Listing 3–7 shows the implementation of the CoinsControllerDelegate protocol in the GameController.m file.

Listing 3–7. *GameController.m (CoinsControllerDelegate Tasks)*

```
//CoinsControllerDelegate tasks
-(void)gameDidStart:(CoinsController*)aCoinsController with:(CoinsGame*)game{
    for (UILabel* label in scoreLabels){
        [label setText:[NSString stringWithFormat:@"%d", [game score]]];
    }
    for (UILabel* label in remainingTurnsLabels){
        [label setText:[NSString stringWithFormat:@"%d", [game remaingTurns]]];
    }
}
-(void)scoreIncreases:(CoinsController*)aCoinsController with:(int)newScore{
    for (UILabel* label in scoreLabels){
        [label setText:[NSString stringWithFormat:@"%d", newScore]];
    }
}
-(void)turnsRemainingDecreased:(CoinsController*)aCoinsController
with:(int)turnsRemaining{
    for (UILabel* label in remainingTurnsLabels){
        [label setText:[NSString stringWithFormat:@"%d", turnsRemaining]];
    }
}
```

```
-(void)gameOver:(CoinsController*)aCoinsController with:(CoinsGame*)game{
    [continueButton setHidden:YES];

    Score* score = [Score score:[game score] At:[NSDate date]];
    [highscoreController addScore:score];

    UIWindow* window = [[[UIApplication sharedApplication] windows] objectAtIndex:0];
    window.rootViewController = nil;
    window.rootViewController = self;

    [UIView beginAnimations:nil context:@"flipTransitionToBack"];
    [UIView setAnimationDuration:1.2];
    [UIView setAnimationTransition:UIViewAnimationTransitionFlipFromRight
forView:self.view cache:YES];

    [coinsController.view removeFromSuperview];
    [self.view addSubview:highscoreController.view];


    [UIView commitAnimations];
}
```

Each implementation of the tasks defined by the CoinsControllerDelegate protocol for
the class GameController is shown in Listing 3–7. When a CoinsController starts a new
game, it calls gameDidStart:with: on the delegate object. In our case, we want to make
sure the remaining turns labels and the score labels reflect the starting values of the
game. These could be just about anything, because we don't know if we are starting a
new game or continuing an old game. Similarly, these labels are updated whenever the
tasks scoreIncreases:with: and turnsRemainingDecreased:with: are called.

Lastly, in Listing 3–7, the tasks gameOver:with: is called when the game is over. In this
task, we record the high score and switch to the High Score view. Because we allow the
game to be played in landscape or portrait orientation, the phone might be in landscape
when the game is over. If that is the case, we want to make sure the interface orientation
is correct, so the High Score view is displayed correctly. One way to do this is to simply
reset our UIWindow's rootViewController property. This causes
shouldAutorotateToInterfaceOrientation: to be called and makes sure our views are
shown in the correct orientation.

HighscoreController: A Simple, Reusable Component

The HighscoreController class is responsible for managing a view for displaying high
scores as well as persisting those high scores. The HighscoreController is a pretty
simple class—the header file for the class is shown in Listing 3–8.

Listing 3–8. *HighscoreController.h*

```
#define KEY_HIGHSCORES @"KEY_HIGHSCORES"

#import <UIKit/UIKit.h>
#import "Highscores.h"
#import "Score.h"
```

```
@interface HighscoreController : UIViewController {
    IBOutlet UIView* highscoresView;
    Highscores* highscores;
}
-(void)saveHighscores;
-(void)layoutScores:(Score*)latestScore;
-(void)addScore:(Score*)newScore;

@end
```

As can be seen in Listing 3–8, HighscoreController has two fields. The first field, highscoresView, is an IBoutlet. This is the UIView that will be used to lay out the actual scores on the screen. The UIView highscoresView is assumed to be a sub-view of the view property that comes with the class UIViewController. It does not have to be a direct sub-view—it just has to appear somewhere in the view hierarchy. This is a slightly different pattern from other UIViewControllers we have seen. It was done this way so an instance of HighscoreController could be added to both the iPhone and iPad XIB files and the layout of the Views could be controlled in there. Inspect the iPhone XIB file and you will see that the layout for the highscore view is defined directly within the HighscoreController.

The field highscores is of type Highscores. This is a simple class that contains an array of Score objects. We will take a closer look at the classes Highscores and Score after looking at the implementation of the three tasks defined in HighscoreController.

HighscoreController Implementation and Layout

Once an application has the views for HighscoreController wired up, the most important task is the addScore: task. This task is called when a game is over and the application should update the High Score view as well as make sure the high score information is stored. Listing 3–9 shows the implementation of addScore:.

Listing 3–9. *HighscoreController.m (addScore:)*

```
-(void)addScore:(Score*)newScore{
    [highscores addScore:newScore];
    [self saveHighscores];
    [self layoutScores: newScore];
}
```

In Listing 3–9, we see that the addScore: task takes a new Score as an argument. The newScore object is passed onto the addScore: task of the object highscores, where it is either discarded if it is not high enough to be a high score or inserted into the array of ten Scores stored in highscores. We also see that the task in Listing 3–9 calls a saveHighscores task and then updates the layout of the views by calling layoutScores:. Let's take a look at how the views are updated before we look at saving the high scores. Listing 3–10 shows the implementation of layoutScores:.

Listing 3–10. *HighscoreController.m (layoutScores:)*

```
-(void)layoutScores:(Score*)latestScore{
    for (UIView* subview in [highscoresView subviews]){
        [subview removeFromSuperview];
    }
    CGRect hvFrame = [highscoresView frame];
    float oneTenthHeight = hvFrame.size.height/10.0;
    float halfWidth = hvFrame.size.width/2.0;

    NSDateFormatter *dateFormat = [[NSDateFormatter alloc] init];
    [dateFormat setDateFormat:@"yyyy-MM-dd"];


    int index = 0;
    for (Score* score in [highscores theScores]){
        CGRect dateFrame = CGRectMake(0, index*oneTenthHeight, halfWidth,
oneTenthHeight);
        UILabel* dateLabel = [[UILabel alloc] initWithFrame:dateFrame];
        [dateLabel setText: [dateFormat stringFromDate:[score date]]];
        [dateLabel setTextAlignment:UITextAlignmentLeft];

        [highscoresView addSubview:dateLabel];

        CGRect scoreFrame = CGRectMake(halfWidth, index*oneTenthHeight, halfWidth,
oneTenthHeight);
        UILabel* scoreLabel = [[UILabel alloc] initWithFrame:scoreFrame];
        [scoreLabel setText:[NSString stringWithFormat:@"%d", [score score]]];
        [scoreLabel setTextAlignment:UITextAlignmentRight];

        [highscoresView addSubview:scoreLabel];

        if (latestScore != nil && latestScore == score){
            [dateLabel setTextColor:[UIColor blueColor]];
            [scoreLabel setTextColor:[UIColor blueColor]];
        } else {
            [dateLabel setTextColor:[UIColor blackColor]];
            [scoreLabel setTextColor:[UIColor blackColor]];
        }

        index++;
    }

}
```

The task layoutScores: from Listing 3–10 takes a Score object as an argument. This
Score object represents the most recent score the player has achieved. This allows
HighscoreController to mark the newest score in blue; the other scores will be drawn in
black. The first loop in layoutScores: simply removes all sub-views from the UIView
highscoreView. The next three lines inspect the size of highscoreView and pre-compute
some size values that will be used later.

The second loop in layoutScores: iterates through all of the Score objects contained in
the highscores object. For each Score, two UILabels are created. The first UILabel,
called dateLabel is created with the CGRect dateFrame, which defines the region where
the UILabel should be drawn. Basically, dateFrame specifies the left half of a row on

highscoreView. The text for dateLabel is set based on the date property of the Score object. Similarly, this process is repeated for the UILabel scoreLabel; however it will display the score the user achieved and will be placed on the right.

Lastly, we check to see if the score we are displaying is the object latestScore. If it is, we adjust the colors of the UILabels to blue.

If we take a look back at Listing 3–9, we see that the high scores are saved before we update the views by calling the task saveHighscores, as shown in Listing 3–11.

Listing 3–11. *HighscoreController.m (saveHighscores)*

```
-(void)saveHighscores{
    NSUserDefaults* defaults = [NSUserDefaults standardUserDefaults];

    NSData* highscoresData = [NSKeyedArchiver archivedDataWithRootObject: highscores];
    [defaults setObject:highscoresData forKey: KEY_HIGHSCORES];
    [defaults synchronize];
}
```

The saveHighscores task, shown in Listing 3–11, is responsible for archiving the highscores object and writing it somewhere permanent. The strategy here is to stick the highscores object into the users preferences for this application. This way, the high scores will be preserved if the users delete the app after a sync with iTunes. To know the user's preferences, we call standardUserDefaults on the class NSUserDefaults. Objects of type NSUserDefaults are basically maps for storing key value pairs. The keys are NSStrings and the values must be a property list. This includes NSData, NSString, NSNumber, NSDate, NSArray, and NSDictionary—basically the core iOS objects types. We want to store an object of type Highscores, which is not included in this list. In order accomplish this, we must create an NSData object from the data stored in the highscores object.

To archive an object to an NSData object, we use the class NSKeyedArchiver and pass the highscores object to the tasks archivedDataWithRootObject:. As the name implies, archivedDataWithRootObject is designed to archive a graph of objects. In our case, the root object is highscores and we know it contains a number of Score objects. So it looks like we are on the right track. In order for an object to be archived by an NSKeyedArchiver, it must conform to the protocol NSCoding. The final step is to call synchronize on defaults; this makes sure our changes are saved.

The Highscores Class

Instances of the class Highscores store a sorted list of Score objects and handle the details of adding Score object to the list. Let's take a look at the header file for the class Highscores and see how this all works, as shown in Listing 3–12.

Listing 3–12. *Highscores.h*

```
#import <Foundation/Foundation.h>
#import "Score.h"

@interface Highscores : NSObject <NSCoding>{
    NSMutableArray* theScores;
```

```
}
@property (nonatomic, retain) NSMutableArray* theScores;

-(id)initWithDefaults;
-(void)addScore:(Score*)newScore;

@end
```

Listing 3–12 shows the interface for the class Highscores. As we can see, Highscores
does in fact conform to the protocol NSCoding. We also see that it contains an
NSMutableArray called theScores, which is accessible as a property. There are two tasks
defined: one for initializing a Highscore with ten default Scores and one for adding a new
Scores object. Listing 3–13 shows how this class is implemented.

Listing 3–13. *Highscores.m*

```
#import "Highscores.h"

@implementation Highscores
@synthesize theScores;

-(id)initWithDefaults{
    self = [super init];
    if (self != nil){
        theScores = [NSMutableArray new];
        for (int i=0;i<10;i++){
            [self addScore:[Score score:1 At:[NSDate date]]];
        }
    }
    return self;
}
-(void)addScore:(Score*)newScore{
    [theScores addObject:newScore];
    [theScores sortUsingSelector:@selector(compare:)];

    while ([theScores count] > 10){
        [theScores removeObjectAtIndex:10];
    }
}

- (void)encodeWithCoder:(NSCoder *)encoder{
    [encoder encodeObject:theScores forKey:@"theScores"];
}
- (id)initWithCoder:(NSCoder *)decoder{
    theScores = [[decoder decodeObjectForKey:@"theScores"] retain];
    return self;
}
-(void)dealloc{
    for (Score* score in theScores){
        [score release];
    }
    [theScores release];
    [super dealloc];
}
@end
```

The implementation of the class `Highscores` shown in Listing 3–13 is pretty compact. The task `initWithDefaults` initializes the `NSMutableArray` `theScore` and then fills `theScores` with ten new `Score` objects. The task `addScore:` adds a new `Score` object to the `theScores`, sorts it by the score achieved by the player, and then removes any extra `Scores`. This may result in the `Score` `newScore` not actually being in the `NSMutableArray` `theScores`. This is implemented in this way so the caller does not have to consider the fact that the `theScore` might not be high enough to be considered an actual high score. The last two tasks, `encodeWithCoder:` and `initWithCoder:`, are from the protocol `NSCoding`. These tasks describe how a `Highscores` object is archived and unarchived. Note that the object passed to both of these arguments is of the same type: `NSCoder`. The class `NSCoder` provides tasks for encoding and decoding values. `NSCoder` is very much like other iOS classes in that it presents a map-like interface for reading and writing data. In the task `encodeWithCoder:`, we use the `NSCoder` task `encodeObject:forKey:` to write `theScore` object to the encoder. We pass in a key value `NSString` that we will use in `initWithCoder:` to read the `theScores` back out when we unarchive this class. Also note that the object that is returned from the `decodeObjectForKey:` task it retained. This is done to make sure the object returned is not reclaimed at some unspecified time.

When an `NSMutableArray` is encoded with an `NSCoder`, it knows to encode the elements in the array, but those elements must know how to be encoded. Because `theScores` is an `NSMutableArray` filled with `Score` objects, we have to tell the class `Score` how to encode and decode itself for this process to work.

The Score Class

`Score` objects represent a date and score value. We have seen how the class `Highscores` manges a list of `Score` obects. Let's take a quick look at this simple class. Listing 3–14 shows the header of the `Score` class.

Listing 3–14. *Score.h*

```
#import <Foundation/Foundation.h>

@interface Score : NSObject  <NSCoding>{
    NSDate* date;
    int score;
}
@property (nonatomic, retain) NSDate* date;
@property (nonatomic) int score;

+(id)score:(int)aScore At:(NSDate*)aDate;
@end
```

As can be seen in Listing 3–14, we see that it defines two properties, date and score. There is also a utility constructor for quickly creating a `Score` object with these two properties filled in. As mentioned, the class `Score` must conform to the `NSCoding` protocol in order for our archiving process to work. Listing 3–15 shows the implementation of the `Score` class.

Listing 3–15. *Score.m*

```
#import "Score.h"

@implementation Score
@synthesize date;
@synthesize score;

+(id)score:(int)aScore At:(NSDate*)aDate{
    Score* highscore = [[Score alloc] init];
    [highscore setScore:aScore];
    [highscore setDate:aDate];
    return highscore;
}
- (void)encodeWithCoder:(NSCoder *)encoder{
    [encoder encodeObject:date forKey:@"date"];
    [encoder encodeInt:score forKey:@"score"];
}
- (id)initWithCoder:(NSCoder *)decoder{
    date = [[decoder decodeObjectForKey:@"date"] retain];
    score = [decoder decodeIntForKey:@"score"];
    return self;
}
- (NSComparisonResult)compare:(id)otherObject {
    Score* otherScore = (Score*)otherObject;
    if (score > [otherScore score]){
        return NSOrderedAscending;
    } else if (score < [otherScore score]){
        return NSOrderedDescending;
    } else {
        return NSOrderedSame;
    }
}
@end
```

Listing 3–15 shows us a couple of things. We see that the implementation of score:At: simply creates a new Score object and populates the properties. The task compare: is used by Highscores to sort the Score objects (see the task addScore from Listing 3–13). Lastly, we see the now familiar task for archiving and unarchiving: encodeWithCoder: and initWithCoder:. In the case of the Score class, we store the NSDate date, with the object methods of NSCoder. For the int score, we have to use special tasks, because an int is not an object. NSCoder provides a special task for primitive types. In our case, we use decodeInt:ForKey: and encodeInt:ForKey:. There exists analog task for the other primitive types, like BOOL, float, double, and so on.

We have looked at the implementation required by a class we wish to archive (and unarchive), but we have not looked at how we actually unarchive an object from the user's preferences. Listing 3–16 shows how this is done in the class HighscoreController.

Listing 3–16. *HighscoreController (viewDidLoad)*

```
- (void)viewDidLoad
{
    [super viewDidLoad];
    [highscores release];

    NSData* highscoresData = [[NSUserDefaults standardUserDefaults]
dataForKey:KEY_HIGHSCORES];

    if (highscoresData == nil){
        highscores = [[[Highscores alloc] initWithDefaults] retain];
        [self saveHighscores];
    } else {
        highscores = [[NSKeyedUnarchiver unarchiveObjectWithData: highscoresData]
retain];
    }
    [self layoutScores:nil];
}
```

The viewDidLoad task from Listing 3–16 is called when a HighscoreController's view is loaded. In this task, we want to prepare the HighscoreController for use. This means making sure highscores is in a usable and accurate state, and we want to lay out the current set of high scores, in case this view is displayed before a new Score is added. In order to retrieve the current high scores, we read a NSData object from the NSUserDefaults object returned by the call to standardUserDefaults. If the NSData object is nil, which will happen the first time you run the application, we initialize a new Highscores object and immediately save it. If the NSData object is not nil, we unarchive it with a call to NSKeyedUnarchiver's task unarchiveObjectWithData and retain the results.

We have looked at how an object can be archived and unarchived in this section. We will now use the same principles to show how we can archive and unarchive our game state.

Preserving Game State

With the advent of background execution on iOS devices, the importance of preserving game or application state is a bit reduced. But a fully functional and user-friendly application should be able to gracefully handle restoring state if the application is terminated. The steps required to preserve state are really not that different from storing other types of data. This is a critical feature of many applications.

The first thing we need to understand is when we should be trying to restore state or archive state. Figure 3–11 is a flowchart describing our sample application's life cycle in terms of initialization.

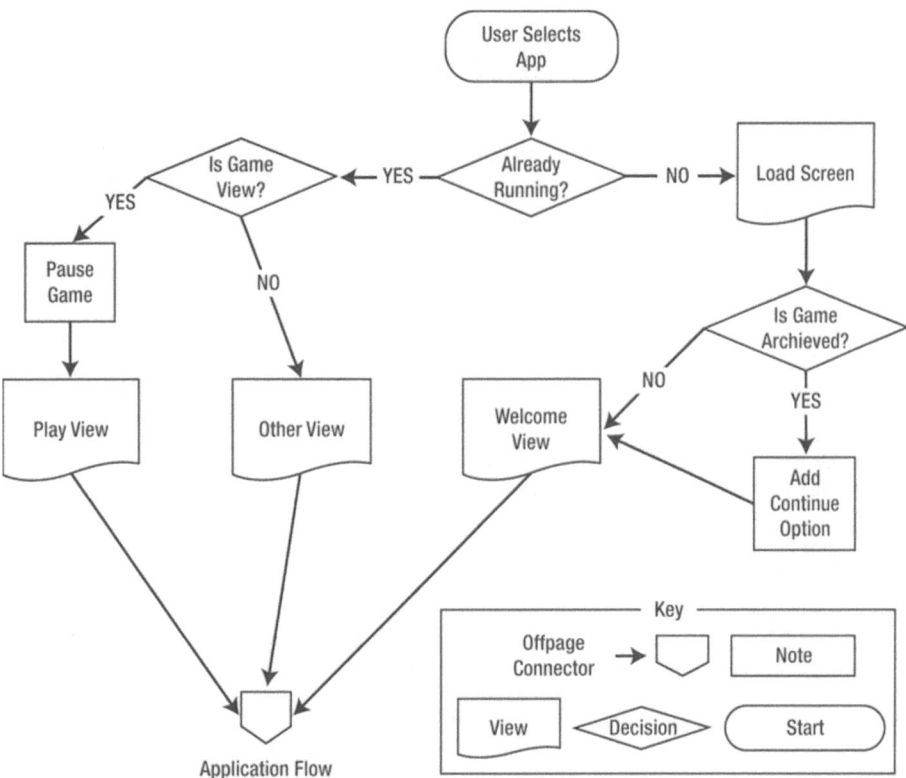

Figure 3–11. *Initialization life cycle of our example application*

In Figure 3–11, we see that all applications start by the user clicking on the application icon. From there, one of two things happen: either the application, if already running in the background, will be brought to the foreground, or the application will start fresh. If the application was already running, it will show up on the screen exactly as the user left it; no code is required on our part to reestablish state. In an action game, you might want to pause the game when the application is brought to the foreground; this will give the user a chance a reorient before play resumes. Either way, we return to the application flow presented in Figure 3–1.

If the application was not already running, we show the load screen as normal, and then we check to see if the game state was archived. If yes, we want to unarchive the game state and show the Continue button. We then continue with the flow presented in Figure 3–1. To support the case where we need to unarchive the game state, we obviously need to archive it at some point. Let's start by looking at the archive logic, and then looking at the unarchive logic.

Archiving and Unarchiving Game State

In our sample application, we store the state of our game in a class called CoinsGame. We are going to save some of the implementation details of this class for the next chapter, but because we know it is being archived and unarchived, we know it probably conforms to the NSCoding protocol. And indeed it does—Listing 3–17 shows the implementation of the two NSCoding tasks.

Listing 3–17. *CoinsGame.m (encodeWithCoder: and initWithCoder:)*

```
- (void)encodeWithCoder:(NSCoder *)encoder{

    [encoder encodeObject:coins forKey:@"coins"];
    [encoder encodeInt:remaingTurns forKey:@"remaingTurns"];
    [encoder encodeInt:score forKey:@"score"];
    [encoder encodeInt:colCount forKey:@"colCount"];
    [encoder encodeInt:rowCount forKey:@"rowCount"];

}
- (id)initWithCoder:(NSCoder *)decoder{

    coins = [[decoder decodeObjectForKey:@"coins"] retain];
    remaingTurns = [decoder decodeIntForKey:@"remaingTurns"];
    score = [decoder decodeIntForKey:@"score"];
    colCount = [decoder decodeIntForKey:@"colCount"];
    rowCount = [decoder decodeIntForKey:@"rowCount"];

    return self;
}
```

In Listing 3–17, we see the task used by CoinsGame to support archiving and unarchiving. In each task, a number of properties are encoded or decoded—nothing to surprising there, I hope. An object can be archived at any point in the application life cycle, but there are special tasks that are called when an application is either closing or terminating. These are the following:

```
- (void)applicationWillTerminate:(UIApplication *)application
- (void)applicationDidEnterBackground:(UIApplication *)application
```

The task applicationWillTerminate: is called just before the application is terminated. Its cousin task applicationDidEnterBackground: is similar, but it is called when the application is sent to the background. If your application supports background execution (the default in new projects), the task applicationWillTerminate: will not be called; you must put your archiving logic in applicationDidEnterBackground:. This is the case because a user can exit an application in the background at any point, so it makes sense to have the bookkeeping taken care of beforehand, while the application is fully active. There are a number of other life cycle tasks available to app delegates. When you create a new project in Xcode, these tasks are automatically added to your app delegate class with nice documentation.

Implementing Life Cycle Tasks

Because our application supports background execution, we put the archiving logic into the applicationDidEnterBackground: task, as shown in Listing 3–18.

Listing 3–18. *Sample_03AppDelegate.m (applicationDidEnterBackground:)*

```
- (void)applicationDidEnterBackground:(UIApplication *)application
{
    NSString* gameArchivePath = [self gameArchivePath];
    [NSKeyedArchiver archiveRootObject:[gameController currentGame] toFile:
gameArchivePath];
}
```

To archive our game state, Listing 3–18 shows that we again use the NSKeyedArchiver class to archive our CoinsGame object, but this time we archive it to a file. The variable gameArchivePath is the path to the file we are going to use as our archive. We get this path by calling the task gameArchivePath, as shown in Listing 3–19.

Listing 3–19. *Sample_03AppDelegate.m (gameArchivePath)*

```
-(NSString*)gameArchivePath{
    NSArray* paths = NSSearchPathForDirectoriesInDomains(NSDocumentDirectory,
NSUserDomainMask, YES);
    NSString* documentDirPath = [paths objectAtIndex:0];
    return [documentDirPath stringByAppendingPathComponent:@"GameArchive"];
}
```

The gameArchivePath task from Listing 3–19 shows that we use the function NSSearchPathForDirectoriesInDomains and pass the NSDocumentDirectory, masked with the NSUserDomainMask. The Yes at the end indicates that we want the tilde that is used to indicate the user's home directory to be expanded into the full path. Each application on an iOS device is given a root directory to which files can be read and written. By getting the zero items out of the path's NSArray, we get access to that directory. We simply specify the name of the file we want to use passing an NSString to the stringByAppendingPathComponent task of documentDirPath.

We obviously need the path to the archive file when we unarchive the CoinsGame object as well. Listing 3–20 shows where we unarchive this object.

Listing 3–20. *Sample_03AppDelegate.m(application: didFinishLaunchingWithOptions:)*

```
- (BOOL)application:(UIApplication *)application
didFinishLaunchingWithOptions:(NSDictionary *)launchOptions
{

    NSString* gameArchivePath = [self gameArchivePath];
    CoinsGame* existingGame;
    @try {
        existingGame = [[NSKeyedUnarchiver unarchiveObjectWithFile:gameArchivePath]
retain];
    }
    @catch (NSException *exception) {
        existingGame = nil;
    }
```

```
    [gameController setPreviousGame:existingGame];
    [existingGame release];

    [self.window makeKeyAndVisible];
    return YES;
}
```

In Listing 3–20, we are looking at the task that is called when our application is fully loaded and ready to go. You will recall the last two lines from the previous chapter: they put the UIWindow on the screen and return that everything is okay. The earlier part of the task is concerned with unarchiving our game state. Using the gameArcivePath, we call unarchiveObjectFromFile on the class NSKeyedUnarchiver. This returns our previously archive CoinsGame object, if it exists. If this is the first time the application was run, existingGame will be nil. Additionally, if an exception occurred, we set existingGame to nil as well, because we know that calling setPreviousGame on gameController can handle a nil value. To complete the cycle, let's take a look at GameController and see what it does with this unarchived CoinsGame in Listing 3–21.

Listing 3–21. *GameController.m(setPreviouseGame:)*

```
-(void)setPreviousGame:(CoinsGame*)aCoinsGame{
    previousGame = [aCoinsGame retain];

    if (previousGame != nil && [previousGame remaingTurns] > 0){
        [continueButton setHidden:NO];
    } else {
        [continueButton setHidden:YES];
    }
}
```

In Listing 3–21, we see that the passed in CoinsGame is saved as the local field previousGame and retained. If previousGame is not nil and there are turns remaining, we display the button continueButton; otherwise, we hide it. Once the user is shown the Welcome view, they will be able to continue playing the game where they left off.

Summary

In this chapter, we explored supporting elements required by a complete game. These included establishing a flow of an application and managing the views that comprise this flow. We looked at a way to coordinate the game behaviors, such as a change in score, or the end of a game with the rest of the application using the delegate pattern. We also looked at two ways of persisting data: through the user settings or by storing an object on disk. Both techniques use the same archiving and unarchiving techniques exposed thought the protocol NSCoding.

Chapter 4

Quickly Build an Input-Driven Game

All games are driven by user input, but depending on how the game behaves between user actions, a game may be one of two types. One type is the action game, in which the events on the screen unfold whether or not the user provides any input. Action games are explored in Chapter 5. In this chapter, we are going to look at games that wait for the user to make a choice. Games of this type include puzzle games and strategy games. In this chapter, we refer to them as *input-driven games*.

Although the coolness factor of input-driven games is definitely less than that of action games, many successful games of this type are on the market. From Minesweeper to Sudoku to Angry Birds, these types of games have captured large audiences, and so it is important to understand how this type of game is implemented. Figure 4–1 shows the typical life cycle of an input-driven game.

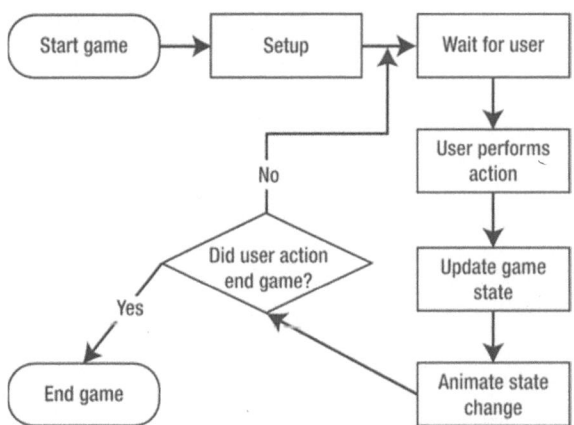

Figure 4–1. *Coin Sorter—a simple input-driven game*

In Figure 4–1, after any initial setup, the application waits for a user to take some action. When the user does something, the game state is updated, and animations are created to reflect the user's action. The animation created by this user action can be simple, such as highlighting a selection, or complex, like an entire physics simulation. After the animation is completed, the application tests to see whether the end of the game has been reached. If it has not, the application once again waits for user input.

In this chapter, you will explore how to create a game that follows the process outlined in Figure 4–1. You will also look at the details of getting content onto the screen and creating animations.

Exploring How to Get Content on the Screen

In the preceding chapter, we assembled views to create a complete application. We used Interface Builder to create UI elements, and for the most part, simply replaced one view with another to move the user from view to view. In this chapter, we are going to take a closer look at the UIView and how it can be used to programmatically create dynamic content. This includes an exploration of how to place components on the screen as well as animate them.

Understanding UIView

The class UIView is the base component class of UIKit. All other UI elements, such as buttons, switches, and scroll views, to name a few, are subclasses of UIView. Even the class UIWindow is a subclass of UIView, so when we manipulate content on the screen, a lot of what we will be doing is interacting with the UIView class.

The UIView class works a lot like the base UI component class in other programming environments. Each UIView defines a region of the screen and has a number of children, or subviews, which are drawn relative to their parent. Figure 4–2 shows an example layout of UIView instances.

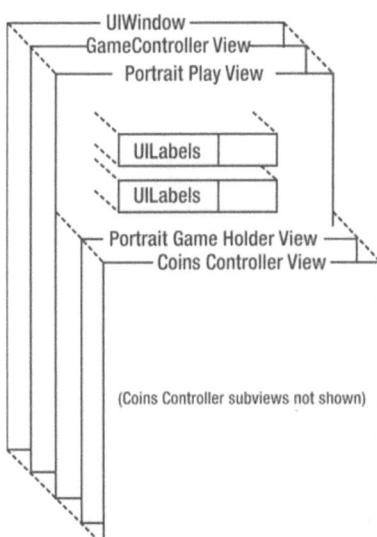

Figure 4–2. *Nested UIView instances*

In Figure 4–2, you can see in the back the root `UIWindow` for our sample application. When the user is seeing the game view, the `UIWindow` has a single subview, which is the view property associated with the `GameController` class. The `GameController` places the Portrait Play view as a subview of its view when the user clicks the New Game button. The Portrait Play view has five subviews, as defined in the XIB file. Four of these subviews are `UILabels` for displaying and tracking the user's score and remaining turns. The fifth `UIView` is the Portrait Game Holder view, which is responsible for defining the region where the view from the `CoinsController` is placed. The view from the `CoinsController` has several subviews, which make up the interactive part of the game. You will be looking at those views in more detail later, but for now, understand that those views are added to the scene in exactly the same way as the rest of the views.

Another thing to notice about Figure 4–2 is that not all of the views are placed in the same position relative to their parent. For example, the Portrait Play view is placed at the upper left of the `GameController` view, while the Portrait Game Holder view is placed about halfway down its parent, the Portrait Play view. The property `frame` of the subview dictates where it is placed relative to its parent. Figure 4–3 shows this in more detail.

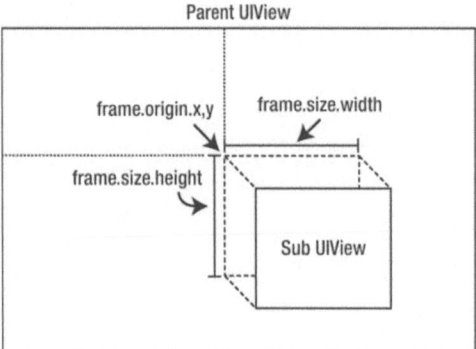

Figure 4–3. *The frame of a UIView*

As shown in Figure 4–3, the frame of a UIView describes not only the location of a subview, but also the size of the view. Generally speaking, the frame is said to describe the region of a subview. The property frame is a struct of type CGRect and is composed of two other structs, origin and size. The field origin is of type CGPoint and describes the location of the upper-left corner of the subview in terms of points. The field size is of type CGSize and describes the number of points the subview is wide and high.

Core Graphics Type Definitions

Core Graphics defines a number of types, structs, and functions. These include the previously mentioned CGRect, CGpoint, and CGSize. Listing 4–1 shows the definitions of these three structs.

Listing 4–1. *CGGeometry.h (CGRect, CGPoint, and CGSize)*

```
/* Points. */

struct CGPoint {
  CGFloat x;
  CGFloat y;
};
typedef struct CGPoint CGPoint;

/* Sizes. */

struct CGSize {
  CGFloat width;
  CGFloat height;
};
typedef struct CGSize CGSize;

/* Rectangles*/
{
  CGPoint origin;
  CGSize size;
};
typedef struct CGRect CGRect;
```

Listing 4–1 presents each of the core structs that define the region of a subview in relation to its parent. These structs are defined in the file CGGeometry.h, which is part of the Core Graphics framework, a standard part of any iOS project. Notice that x, y, width, and height values are defined as CGFloat. The unit for these values is points, not pixels. The difference is subtle at first, but the idea here is that the coordinates and sizes you are specifying with these values is meant to be resolution independent. Further discussion of points vs. pixels can be found in Appendix A.

> **TIP:** The letters *CG* in the names of structs such as CGRect refer to the Core Graphics framework. So a CGRect is referred to as a *Core Graphics Rect* when you want to be precise.

To create any of the base geometric types from Listing 4–1, an inline utility function is defined in CGGeometry.h, as seen in Listing 4–2.

Listing 4–2. *CGGeometry.h (CGPointMake, CGSizeMake, and CGRectMake)*

```
/*** Definitions of inline functions. ***/

CG_INLINE CGPoint
CGPointMake(CGFloat x, CGFloat y)
{
  CGPoint p; p.x = x; p.y = y; return p;
}

CG_INLINE CGSize
CGSizeMake(CGFloat width, CGFloat height)
{
  CGSize size; size.width = width; size.height = height; return size;
}

CG_INLINE CGRect
CGRectMake(CGFloat x, CGFloat y, CGFloat width, CGFloat height)
{
  CGRect rect;
  rect.origin.x = x; rect.origin.y = y;
  rect.size.width = width; rect.size.height = height;
  return rect;
}
```

Listing 4–2 shows the utility functions that can be used to create new geometry values. Only the definition is shown, because the declaration of these functions is trivial. The type CGFloat is defined simply as float. Presumably, this could be changed to support a different architecture. Also, CG_INLINE is simply defined as static inline, indicating that the compiler may inject a compiled version into the calling code.

Using Core Graphics Types

The types and functions defined by Core Graphics are pretty staightforward. Let's look at an example to illustrate their use. Listing 4–3 shows one UIView being added to another at a specific region.

Listing 4–3. *Example of Adding a Subview and Specifying the Frame*

```
UIView* parentView = [[UIView alloc] initWithFrame:CGRectMake(0, 0, 480, 320)];
UIView* subView = [UIView new];

CGRect frame = CGRectMake(200, 100, 30, 40);
[subView setFrame:frame];

[parentView addSubview:subView];
```

In Listing 4–3, we create two UIViews in two different ways. The UIView parentView is created in the more standard of the two ways, by calling alloc and then calling initWithFrame: and passing in a new CGRect to specify its location and size. Another perfectly valid way to create a UIView is by simply calling new and then setting the frame property, as we do for subView. The last thing done in this code snippet is to add subView to parentView by calling addSubview:. The result would be something like Figure 4–3 shown earlier.

Now that you understand the basics of the UIView subclasses and how they are nested and placed, you can take a look at a few basic animations. This will give you the understanding required to explore our simple game.

Understanding Animations

There are two key components to creating an event-driven game. The first is creating animations, and the second is managing the game state. There are several ways to create animations in an iOS application. In the preceding chapter, we implemented animations to transition from view to view. In this section, we will review how that was done and show a similar example to flesh out your understanding. At the end of this chapter, you will take a look at yet a third technique that explores the backing classes used by the first two examples.

The Static Animation Tasks of UIView

The class UIView has static tasks designed to simplify the creation of basic animations. These same tasks can be used to trigger canned animations, such as the transitions used to switch from one view to another. Figure 4–4 shows the canned animation that occurs when the user switches from the High Score view to the Welcome view.

Figure 4–4. *Flip animation*

Starting at the upper left of Figure 4–4, the High Score view flips from left to right until it becomes just a sliver. The animation continues, revealing the Welcome view in a smooth, visually appealing transition. This is just one of the many built-in animations in iOS. Users have learned that this standard transition indicates a change in context. To understand how this animation was triggered, let's start by looking at the steps preformed in Chapter 3 to switch from the High Score view to the Welcome view. These steps are shown in Listing 4–4.

Listing 4–4. *GameController.m (highscoreDoneClicked:)*

```
- (IBAction)highscoreDoneClicked:(id)sender {
    [UIView beginAnimations:@"AnimationId" context:@"flipTransitionToBack"];
    [UIView setAnimationDuration:1.0];
    [UIView setAnimationTransition:UIViewAnimationTransitionFlipFromLeft
forView:self.view cache:YES];

    [highscoreController.view removeFromSuperview];
    [self.view addSubview: welcomeView];


    [UIView commitAnimations];
}
```

Listing 4–4 shows the task called when the user touches the Done button in the High Score view. The first line of this task calls the static task beginAnimations:context:. This task is sort of like starting a database transaction, in that it is creating an animation object, and any change that occurs to UIView instances before commitAnimations is called is part of this animation. The first NSString passed to beginAnimations:context: is an identifier for the backing animation object. The second NSString is similar to the first in that it is used to give additional context about when the animation was created. These two strings could be nil in this example, because we don't really care much about this animation after we define it. In the next example, we will take advantage of these strings to define code to be run when the animation is done.

After the call to beginAnimations:context:, the duration of the animation is defined by calling setAnimationDuration:. The value passed in is the number of seconds the animation should run. The call to setAnimationTransition:forView:cache: indicates that we want to use one of the canned animations that comes with iOS. The animation we have selected is UIViewAnimationFlipFromLeftForView. Because this is a transition animation, we specify the view self.view, which is the super view whose content is changing.

After the animation is set up, the next thing we do (as shown in Listing 4–4) is make the changes to our scene that we want animated. Because this task is responsible for switching the displayed view from the High Score view to the Welcome view, we simply remove the view highscoreController.view from its super view (self.vew) and add the welcomeView to self.view. Finally, we commit the animation, causing it to be displayed to the user.

We can use these static tasks of UIView to create other animations, not just the ones provided by iOS. Let's update the transition from the Welcome screen to the High Score screen so one fades into the other, as seen in Figure 4–5.

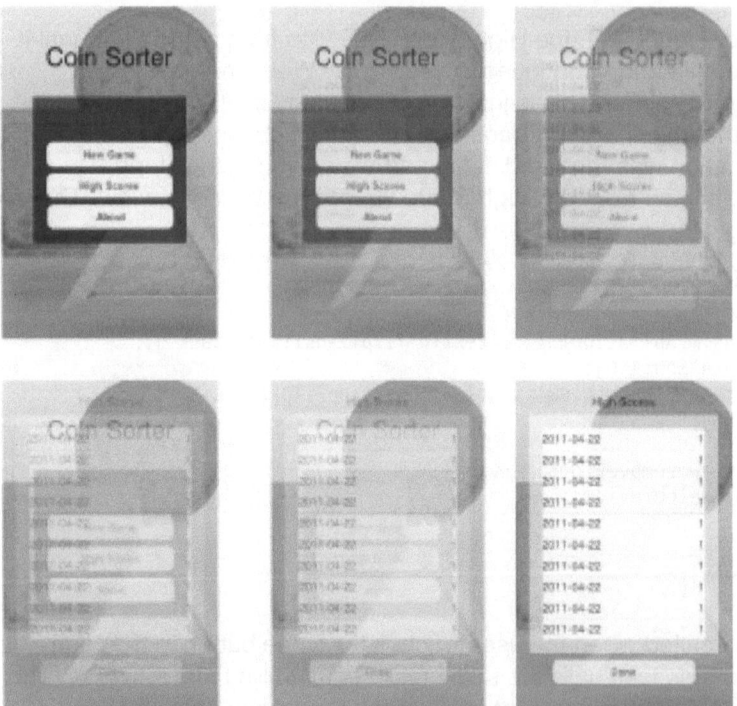

Figure 4–5. *Fade transition animation*

In this fade transition, the Welcome view is replaced with the High Score view. This is accomplished by changing the opacity of each view over a period of time. During this animation, both views are part of the scene. At the end, we want to remove the Welcome view, to stay consistent with how we are managing views in the game. The code that creates this transition is shown in Listing 4–5.

Listing 4–5. *GameController.m (highScoresClicked:)*

```
- (IBAction)highScoresClicked:(id)sender {

    //Make high score view 100% clear and add it on top of the welcome view.
    [highscoreController.view setAlpha:0.0];
    [self.view addSubview:highscoreController.view];

    //set up animation
    [UIView beginAnimations:nil context:@"fadeInHighScoreView"];
    [UIView setAnimationDuration:1.0];
    [UIView setAnimationDelegate:self];
    [UIView setAnimationDidStopSelector:@selector(animationDidStop:finished:context:)];

    //make changes
    [welcomeView setAlpha:0.0];
    [highscoreController.view setAlpha:1.0];

    [UIView commitAnimations];
}
```

The task shown in Listing 4–5 is called when the user touches the Highscores button in the Home view. To create the fade effect, we utilize the static animation task from the class UIView, but first we have to set up a few things. We start by setting the alpha property of the view associated with the highscoreController to 0.0, or completely translucent. We then add the High Score view to the root view. This places it onto the already present welcomeView, but because it is clear, we see only the Welcome screen.

The animation is set up by calling the task beginAnimations:context:, and the duration is set with the task setAnimationDuration:. The next step is to set a delegate for the animation that is being created. We also want to set which task should be called when the animation is complete by calling setAnimationDidStopSelector:. After the animation is set up, we simply set the alpha property of welcomeView to 0.0 and set the High Score view's alpha to 1.0. This indicates that at the end of the animation, we want the Welcome view to be transparent and the High Score view to be fully opaque. We then indicate that we have set our finished state by calling commitAnimations.

We set self to be the delegate for the animation and set up the task animationDidStop:finished:context: to be called when we want the animation to be over. We did this so we could do a little cleanup at the end of the animation and ensure that our application is in the correct state. Listing 4–6 shows the implementation of this callback method.

Listing 4–6. *GameController.m (animationDidStop:finished:context:)*

```
- (void)animationDidStop:(NSString *)animationID finished:(NSNumber *)finished
context:(void *)context{
    if ([@"fadeInHighScoreView" isEqual:context]){
        [welcomeView removeFromSuperview];
        [welcomeView setAlpha:1.0];
    }
}
```

The task in Listing 4–6 is called at the end of the fade animation. In this task, we want to remove welcomeView from the scene and reset its alpha to 1.0. We remove it, because that is the expected state for the other transitions. Setting the alpha back to 1.0 ensures that welcomeView is visible when it is added back into the scene.

You have seen how a view is added to a super view by using the property frame. You have also learned how to create animations that affect the properties of a UIView by exploring the fade transition. We are going to combine these concepts to show you how to animate the coins within the game, but first you have to take look at how we set up the related classes so that this all makes sense.

Building the Game Coin Sorter

Thus far we have looked at some of the finer points of getting content onto the screen. In previous chapters, we explored the role of the controller classes and how they manage the relationship between data and the view. The last step in creating the game Coin Sorter is combining these concepts. This section will walk through the life cycle of the CoinController class, touching on the basic concepts we have explored. Figure 4–6 shows just the area of the screen controlled by the class CoinController.

In this five-by-five grid of coins, the user can select two coins to trade places. The goal is to create rows or columns of like coins to create a matching set. When a match is created, the coins are animated on the screen, and the user's score is increased. The user has 10 turns to make as many matches as possible. Figure 4–7 shows the life cycle of the game.

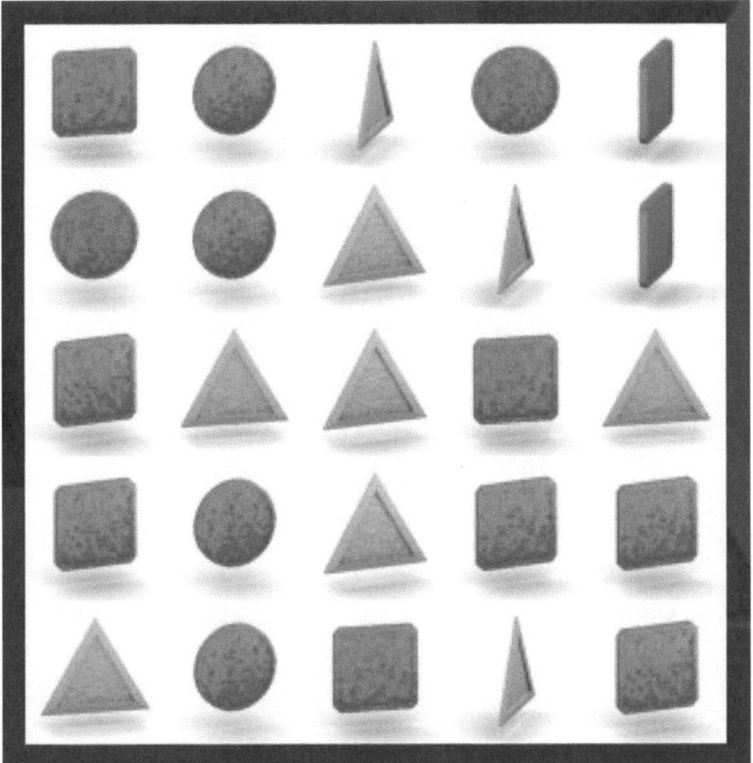

Figure 4–6. *CoinController's view*

Figure 4–7 shows the flow of the application for a single game. After setup, the application waits for the user to select a coin. If a first coin is not selected, the application simply keeps track that the coin was selected and goes back to waiting for the user. When the user selects a coin a second time, the application checks whether the coin just selected is the same coin as the first one selected. If so, the application unselects the coin. This allows the user to change his mind. If the user selects a different coin, the game state updates by modifying the CoinGame object, and animations are created, showing the two coins trading places. When the coin-swapping animation is done, the application checks whether any matches exist. If so, it animates their removal and updates CoinGame with new coins to replace those just removed. Because adding new coins could create additional matches, the application checks again for matches. This process could theoretically go on forever, but in practice it does not. Eventually, there will be no matches to remove.

When there are no matches, the application checks whether the user has any turns remaining. If the user does, it goes back to waiting for the user's input. If the user does not, the game is over.

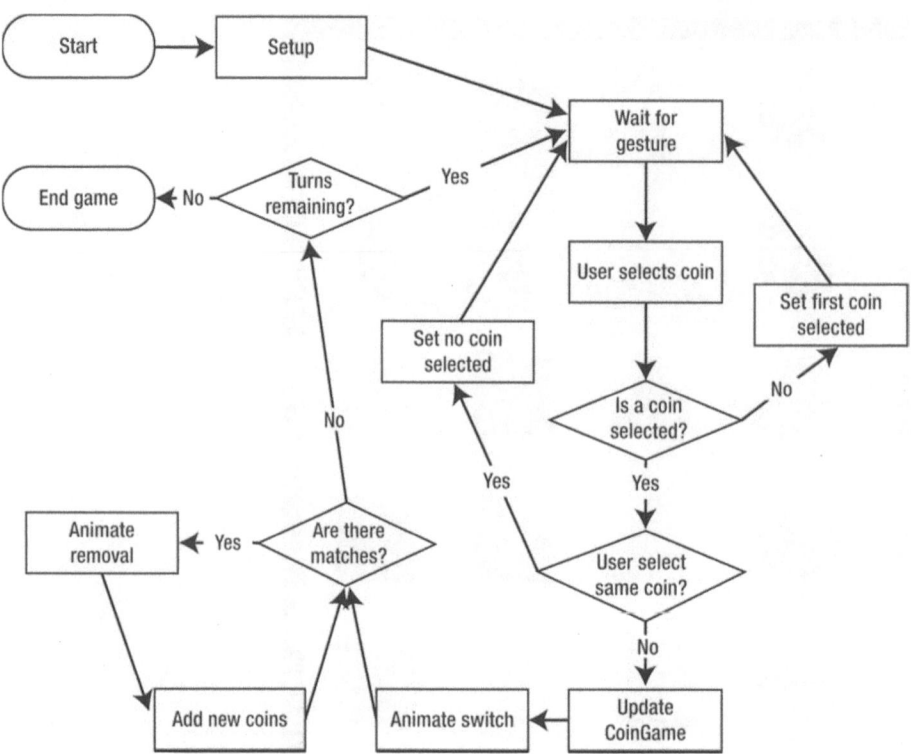

Figure 4–7. *Coin Sorter life cycle*

Implementing Game State

The logic presented in Figure 4–7 is implemented in the class CoinsController. To help you understand how the class CoinsController creates the Coin Sorter game, let's take a look at the header file. There you can get an overview of the class, shown in Listing 4–7.

Listing 4–7. *CoinsController.h*

```
#import <UIKit/UIKit.h>
#import "CoinsGame.h"

@class CoinsController;
@protocol CoinsControllerDelegate <NSObject>

-(void)gameDidStart:(CoinsController*)aCoinsController with:(CoinsGame*)game;
-(void)scoreIncreases:(CoinsController*)aCoinsController with:(int)newScore;
-(void)turnsRemainingDecreased:(CoinsController*)aCoinsController
with:(int)turnsRemaining;
-(void)gameOver:(CoinsController*)aCoinsController with:(CoinsGame*)game;

@end
```

```objc
@interface CoinsController : UIViewController {
    CoinsGame* coinsGame;
    UIView* coinsView;

    NSMutableArray* imageSequences;

    BOOL isFirstCoinSelected;
    Coord firstSelectedCoin;
    Coord secondSelectedCoin;
    BOOL acceptingInput;

    UIImageView* coinViewA;
    UIImageView* coinViewB;
    NSMutableArray* matchingRows;
    NSMutableArray* matchingCols;

    IBOutlet id<CoinsControllerDelegate> delegate;
}
@property (nonatomic, retain) CoinsGame* coinsGame;
@property (nonatomic, retain) id<CoinsControllerDelegate> delegate;

+(NSArray*)fillImageArray:(int)coin;

-(void)loadImages;
-(void)newGame;
-(void)continueGame:(CoinsGame*)aCoinsGame;
-(void)createAndLayoutImages;
-(void)tapGesture:(UIGestureRecognizer *)gestureRecognizer;
-(Coord)coordFromLocation:(CGPoint) location;
-(void)doSwitch:(Coord)coordA With:(Coord)coordB;
-(void)checkMatches;
-(void)updateCoinViews;

-(void)spinCoinAt:(Coord)coord;
-(void)stopCoinAt:(Coord)coord;

-(void)doEndGame;

@end
```

This header file for the class CoinsController describes the class as well as the delegate protocol CoinsControllerDelegate, discussed in Chapter 3. The field CoinsGame is used to store the state of the game. This object can be archived to save the game state or it can passed to an instance of CoinsController to resume a game. The UIView coinsView is the white play area. The BOOL isFirstCoinSelected is used to keep track of whether the user has already selected a coin. The two Coord structs record which coins are selected. We will take a look at the Coord struct shortly. The BOOL acceptingInput is used to block user input when animations are happening.

The two UIImageView fields, coinViewA and coinViewB, along with the NSMutableArray variables matchingRows and matchingCols are used during animations to keep track of what is being animated. The last field is the delegate for this class and must implement the protocol CoinsControllerDelegate.

Initialization and Setup

The life of a CoinController object starts when the task viewDidLoad is called. This is called automatically because we have an instance of CoinController in our XIB files. Listing 4–8 shows this task.

Listing 4–8. *CoinController.m (viewDidLoad)*

```
- (void)viewDidLoad
{
    [super viewDidLoad];

    [self.view setBackgroundColor:[UIColor clearColor]];

    CGRect viewFrame = [self.view frame];

    //border is 3.125%
    float border = viewFrame.size.width*.03125;
    float coinsViewWidth = viewFrame.size.width-border*2;
    CGRect coinsFrame = CGRectMake(border, border, coinsViewWidth, coinsViewWidth);

    coinsView = [[UIView alloc] initWithFrame: coinsFrame];
    [coinsView setBackgroundColor:[UIColor whiteColor]];
    [coinsView setClipsToBounds:YES];

    [self.view addSubview:coinsView];

    UITapGestureRecognizer* tapRecognizer = [[UITapGestureRecognizer alloc]
initWithTarget:self action:@selector(tapGesture:)];
    [tapRecognizer setNumberOfTapsRequired:1];
    [tapRecognizer setNumberOfTouchesRequired:1];

    [coinsView addGestureRecognizer:tapRecognizer];

}
```

In the viewDidLoad task, we have several small items to attend to. First, we call the super implementation of viewDidLoad. This is not strictly necessary but is just good practice. If we ever wanted to change the super class of CoinsController, we might spend a lot of time figuring out why things are not initializing correctly.

The next part of the task viewDidLoad is to set up some UIView instances to do what we want. Looking at Figure 4–6, we see that there is a partially transparent border around the white area containing the coins. This partially transparent view is actually the holder view, so to make sure this is visible, we set the root view for this class to be completely transparent.

The white square area is the UIView called coinsView, which is the parent view for the views that will be the coins. We calculate the size of coinsView based on the frame of the parent view. On the iPhone, the value of coinsViewWidth turns out to be 300 points; on the iPad, coinsViewWidth is 720 points. After setting the background of coinsView to white, we set the property clipToBounds to be true. This will prevent the coins from being drawn outside the white area when they are animated offscreen.

The last order of business in viewDidLoad is to register a gesture recognizer to the coinsView. In this case, we want to know anytime the user taps on the coinsView, so we use a UITapGestureRecognizer called tapRecognizer. When tapRecognizer is initialized, we set self as the target and specify the task tapGesture: as the task to be called when tapRecognizer detects the tap gesture. By setting the number of required taps and the number of required touches, we make sure we pick up the correct gestures that the user makes. Adding tapRecognizer to coinsView finishes setting up our gesture recognizer. More information about how gestures work on iOS can be found in Chapter 8.

Starting a New Game

After viewDidLoad is called, the CoinsController will be used either to play a new game or to continue an old game. Let's take a look at the task newGame first, as shown in Listing 4–9.

Listing 4–9. *CoinsController.m (newGame)*

```
-(void)newGame{
    for (UIView* view in [coinsView subviews]){
        [view removeFromSuperview];
    }

    [coinsGame release];
    coinsGame = [[CoinsGame alloc] initRandomWithRows:5 Cols:5];

    [self createAndLayoutImages];
    [delegate gameDidStart:self with: coinsGame];

    acceptingInput = YES;
}
```

The task newGame is called when the New Game button is clicked on the Welcome screen. The first thing to do is remove the old subviews from coinsView. (There may not be any subviews to remove if the user just launched the application.) The object coinsGame is released before being set to a new instance prepopulated with random coin values. The next step is to call createAndLayoutImages, which will place the coins on the screen. This task is called by both continueGame: and newGame, so we will look at it after looking at continueGame:. The last thing to do is inform any delegate that a new game has started and set acceptingInput to YES.

Continuing a Game

If a game was in progress when a user last quit, that user may wish to continue the game from where it left off. This is done by clicking the Continue button in the Welcome view. When that button is pressed, the task continueGame is called, which is pretty similar to newGame. Listing 4–10 shows continueGame:.

Listing 4–10. *CoinsController.m (continueGame:)*

```
-(void)continueGame:(CoinsGame*)aCoinsGame{
    for (UIView* view in [coinsView subviews]){
        [view removeFromSuperview];
    }

    [coinsGame release];
    coinsGame = aCoinsGame;

    [self createAndLayoutImages];
    [delegate gameDidStart:self with: coinsGame];

    acceptingInput = YES;
}
```

The task continueGame: takes a CoinsGame object as an argument called aCoinsGame. After clearing away the old subviews of coinsView, we set coinsGame to the passed-in aCoinsGame object. A call to createAndLayoutImages is made, to add the coins to the scene. We call the delegate task gameDidStart:coinsGame: so the delegate has a chance to update the UILabel objects that track the current score and remaining turn. Finally, we set acceptingInput to YES.

Initializing the UIViews for Each Coin

As mentioned, the task createAndLayoutImages is called by both newGame and continueGame:. This task is responsible for the initial setup of the game, primarily creating a UIImageView for each coin in the game. This task is shown in Listing 4–11.

Listing 4–11. *CoinsController.m (createAndLayoutImages)*

```
-(void)createAndLayoutImages{
    int rowCount = [coinsGame rowCount];
    int colCount = [coinsGame colCount];

    CGRect coinsFrame = [coinsView frame];
    float width = coinsFrame.size.width/colCount;
    float height = coinsFrame.size.height/rowCount;


    for (int r=0;r<rowCount;r++){
        for (int c=0;c<colCount;c++){

            UIImageView* imageView = [[UIImageView alloc] init];
            CGRect frame = CGRectMake(c*width, r*height, width, height);
            [imageView setFrame:frame];

            [coinsView addSubview: imageView];
            [self spinCoinAt:CoordMake(r, c)];
        }
    }
}
```

Listing 4–11 shows the number of rows and columns to be used in the game. In this example, these are both five. The frame of coinsView is stored in the variable coinsFrame for easy reference, and the width and height of each coin view is calculated. The two for loops create an imageView, set its frame, and add it to coinsView for each row and column. In the nested loops, spinCoinAt: is called, which creates the spinning coin effect. Note that the argument being passed to spinCoinAt: is the result of the function CoordMake. We will take a look at this function and the class CoinsGame after looking at spinCoinAt:, shown in Listing 4–12.

Listing 4–12. *CoinsController.m (spinCoinAt:)*

```
-(void)spinCoinAt:(Coord)coord{
    UIImageView* coinView = [[coinsView subviews] objectAtIndex:[coinsGame
indexForCoord:coord]];

    NSNumber* coin = [coinsGame coinForCoord:coord];
    NSArray* images = [imageSequences objectAtIndex:[coin intValue]];
    [coinView setAnimationImages: images];
    NSTimeInterval interval = (random()%4)/10.0+.6;
    [coinView setAnimationDuration: interval];
    [coinView startAnimating];
}
```

You can see that spinCoinAt: takes an argument of type Coord. This is defined in CoinsGame.h and is a struct that represents a coin at a particular row and column. In the first line of this task, the struct Coord is used to find the index of the UIView that represents this particular coin. This works because there is exactly one UIImageView for each coin. (This was set up in Listing 4–11.) After the correct UIImageView is found, we get the value of the coin from the coinsGame object. The NSNumber coin represents the type of coin for this particular set of coordinates—either a triangle, square, or circle. Using the intValue of the coin, we pull out an NSArray from imageSequences called images. The NSArray image stores all of the images that make up the spinning coin animation. By calling setAnimationImages: on the UIImageView coinView, we indicate that we want this UIImageView to cycle though each UIImage in images to produce the animation effect. Setting the animation duration to a random value and calling startAnimating creates the spinning effect. At this point in the setup, the application will look very much like Figure 4–6 shown earlier. We are ready to start accepting user input. Before we look at that, we should take a closer look at the class CoinsGame so you can understand how we are representing these 25 coins and their types.

The Model

You have now looked at the setup code, which used UIViews and UIImageViews to create a scene on the screen that represents the game. You know that the game is composed of 25 coins in a five-by-five grid. Now we should take a look at the class that is responsible for managing this data, so let's look at CoinsGame.h, shown Listing 4–13.

Listing 4–13. *CoinsGame.h*

```objc
#import <Foundation/Foundation.h>

#define COIN_TRIANGLE 0
#define COIN_SQUARE 1
#define COIN_CIRCLE 2


struct Coord {
    int row;
    int col;
};
typedef struct Coord Coord;

CG_INLINE Coord
CoordMake(int r, int c)
{
    Coord coord;
    coord.row = r;
    coord.col = c;
    return coord;
}

CG_INLINE BOOL
CoordEqual(Coord a, Coord b)
{
    return a.col == b.col && a.row == b.row;
}


@interface CoinsGame : NSObject <NSCoding>{
    NSMutableArray* coins;
    int remaingTurns;
    int score;
    int colCount;
    int rowCount;
}
@property (nonatomic, retain)  NSMutableArray* coins;
@property (nonatomic) int remaingTurns;
@property (nonatomic) int score;
@property (nonatomic) int colCount;
@property (nonatomic) int rowCount;

-(id)initRandomWithRows:(int)rows Cols:(int)cols;

-(NSNumber*)coinForCoord:(Coord)coord;
-(int)indexForCoord:(Coord)coord;

-(void)swap:(Coord)coordA With:(Coord)coordB;
-(NSMutableArray*)findMatchingRows;
-(NSMutableArray*)findMatchingCols;
-(void)randomizeRows:(NSMutableArray*)matchingRows;
-(void)randomizeCols:(NSMutableArray*)matchingCols;


@end
```

Listing 4–13 shows the header file for the class CoinsGame. At the very top of the class, we define three constants: COIN_TRIANGLE, COIN_SQUARE, and COIN_CIRCLE. These values represent the three types of coins found in the game. We also define a struct called Coord that is used to store a row/column value pair. The struct Coord was used earlier in Listing 4–12 to identify a specific coin. There are also two functions to go with the struct Coord. The first, called CoordMake, is used to create a Coord with the corresponding row and column values. The second, called CoordEqual, is used to evaluate whether two Coords are referring to the same coin.

The interface declaration for the class CoinsGame, shown in Listing 4–13, conforms to the protocol NSCoding, so we know we can archive and unarchive instances of this class. We also see that an NSMutableArray called coins is defined. The coins object is used to store the NSNumber values that represent the coins at each coordinate. In addition, ints are used to keep track of the number of remaining turns, the score, and the number of rows and columns used in this game.

There are various tasks defined for the class CoinsGame, and hopefully some of these make sense already. We know we will need a way to swap two coins, so the use of the swap:With: task should make sense. There are also two tasks, findMatchingRows and findMatchingCols, that are used to determine whetherany matches exist. The tasks randomizeRows and randomizeCols are used to set new coin values after a match has been found. Notice that findMatchingRows returns an NSMutableArray and that randomizeRows takes an NSMutableArray. The idea here is that after a match is found and the animations are all done, we can use the same NSMutableArray that represented the match to indicate which coins should be randomized.

Listing 4–13 also has two tasks that take a Coord as an argument. The first one, coinForCoord:, takes a Coord and returns NSNumber, indicating the coin type for that coordinate. The second task, indexForCoord, is used to translate a Coord into int suitable as an index into an array. This task is used to find the coin NSNumber in coinForCoord, and it is also used by CoinsController to find the correct UIImageView for a coin.

Let's take a look at the implementation of some of the tasks, because they will be called by CoinsController at different points. We'll start with initRandomWithRow:Col:, shown in Listing 4–14.

Listing 4–14. *CoinsGame.m (initRandomWithRows:Cols:)*

```
-(id)initRandomWithR.ows:(int)rows Cols:(int)cols{
    self = [super init];
    if (self != nil){
        coins = [NSMutableArray new];

        colCount = cols;
        rowCount = rows;

        int numberOfCoins = colCount*rowCount;

        for (int i=0;i<numberOfCoins;i++){
            int result = arc4random()%3;
            [coins addObject:[NSNumber numberWithInt:result]];
```

```
        }

        //Ensure we don't start with any matching rows and cols.
        NSMutableArray* matchingRows = [self findMatchingRows];
        NSMutableArray* matchingCols = [self findMatchingCols];
        while ([matchingCols count] > 0 || [matchingRows count] > 0){
            [self randomizeRows: matchingRows];
            [self randomizeCols: matchingCols];

            matchingRows = [self findMatchingRows];
            matchingCols = [self findMatchingCols];
        }

        remaingTurns = 10;
        score = 0;
    }
    return self;
}
```

In Listing 4–14, we see the initializer task used to create a new CoinsGame object with randomized coins. After creating a new NSMutableArray and setting it to the variable coins and populate it with the total number of coins we will have in the game. The value of the result will be 0, 1, or 2. After we set up the first set of random values, we have to make sure we don't start the game with any matches. This is done by first finding any matched by calling findMatchingRows and findMatchingCols and seeing whether they contain anything. If they do, we randomize those matches until no more matches are found. Finally, we set the number of remaining turns to 10 and make sure the score starts out at 0. Continuing our exploration of the class CoinsGame, let's look at coinForCoord:, shown in Listing 4–15.

Listing 4–15. *CoinsGame.m (coinForCoord:)*

```
-(NSNumber*)coinForCoord:(Coord)coord{
    int index = [self indexForCoord:coord];
    return [coins objectAtIndex:index];
}
```

The task coinForCoord: takes a Coord and returns the type of coin at that location. Finding the NSNumber object at the given index in the NSMutableArray coins does this. The index is determined by calling indexForCoord:, shown in Listing 4–16.

Listing 4–16. *CoinsGame.m (indexForCoord:)*

```
-(int)indexForCoord:(Coord)coord{
    return coord.row*colCount + coord.col;
}
```

The task indexForCoord: takes a Coord struct. This simple task multiplies the number of columns in the game by the row value of coord and adds the col value.

There are a few more tasks in CoinsGame.m that should be explored to help you understand what CoinsController is doing during different parts of its life cycle. Let's continue by looking at the task swap:With:, shown in Listing 4–17.

Listing 4–17. *CoinsGame.m (swap:With:)*

```
-(void)swap:(Coord)coordA With:(Coord)coordB{
    int indexA = [self indexForCoord:coordA];
    int indexB = [self indexForCoord:coordB];

    NSNumber* coinA = [coins objectAtIndex:indexA];
    NSNumber* coinB = [coins objectAtIndex:indexB];

    [coins replaceObjectAtIndex:indexA withObject:coinB];
    [coins replaceObjectAtIndex:indexB withObject:coinA];

}
```

The task `swap:With:` takes two Coords, `coordA` and `coordB`. This task switches the type of coins at these two coordinates by finding the index of each coordinate, the current value of that index in coins, and switching them. We know that after coins are swapped, `CoinsController` will have to look for any matches. This is done with the tasks `findMatchingRows` and `findMatchingCols`. Listing 4–18 shows `findMatchingRows`.

Listing 4–18. *CoinsGame.m (findMatchingRows)*

```
-(NSMutableArray*)findMatchingRows{
    NSMutableArray* matchingRows = [NSMutableArray new];

    for (int r=0;r<rowCount;r++){
        NSNumber* coin0 = [self coinForCoord:CoordMake(r, 0)];
        BOOL mismatch = false;

        for (int c=1;c<colCount;c++){
            NSNumber* coinN = [self coinForCoord:CoordMake(r,c)];
            if (![coin0 isEqual:coinN]){
                mismatch = true;
                break;
            }
        }
        if (!mismatch){
            [matchingRows addObject:[NSNumber numberWithInt:r]];
        }
    }
    return matchingRows;
}
```

The task `findMatchingRows` creates a new NSMutableArray called `matchingRows` to store any results. Matching rows are found by looking at each row in turn and inspecting the coin at each column. This is done by storing the coin at column 0 as the variable `coin0` and then by comparing `coin0` to each coin in the other columns. If a coin is found that does not match `coin0`, we know that there is no match, and `mismatch` is set to true. If no mismatches are found, we add the value r to the NSMutableArray `matchingRows`, which is the result of this task. The implementation for `findMatchingCols` is trivially different and is omitted for brevity.

We also know that after matching rows are found, `CoinsController` will have to randomize any matches to prepare for the next turn the user takes. This is done with the tasks `randomizeRows:` and `randomizeCols:`. The task `randomizeRows:` is shown in Listing 4–19.

Listing 4–19. *CoinsGame.m (randomizeRows:)*

```
-(void)randomizeRows:(NSMutableArray*)matchingRows{
    for (NSNumber* row in matchingRows){
        for (int c=0;c<colCount;c++){
            int index = [self indexForCoord:CoordMake([row intValue], c)];
            int newCoin = arc4random()%3;
            [coins replaceObjectAtIndex:index withObject:[NSNumber
numberWithInt:newCoin]];
        }
    }
}
```

The task randomizeRows takes an NSMutableArray filled with NSNumbers. Each NSNumber is a row. By iterating over each row, we set a new random value for every coin in that row. The implementation of randomzeCols: is similar; we simply randomize each coin for the columns passed in.

Now that you have an understanding of the class CoinsGame, you can take a further look at CoinsController and see how this class used the data stored in CoinsGame to interpret user input, manage the views representing the coins, and create animations based on game state.

Interpreting User Input

When the user touches one of the coins, we want to mark it as selected. Looking back at Listing 4–8, we know that a UITapGestureRecognizer was created and added to the UIView coinsView. This UITapGestureRecognizer calls the task tapGesture: whenever it registers a tap gesture. Let's look at the first part of tapGesture: in Listing 4–20.

Listing 4–20. *CoinsController.m (tapGesture:, partial)*

```
- (void)tapGesture:(UIGestureRecognizer *)gestureRecognizer{

    if ([coinsGame remaingTurns] > 0 && acceptingInput){
        UITapGestureRecognizer* tapRegognizer =
(UITapGestureRecognizer*)gestureRecognizer;
        CGPoint location = [tapRegognizer locationInView:coinsView];
        Coord coinCoord = [self coordFromLocation:location];

        if (!isFirstCoinSelected){//first of the pair
            isFirstCoinSelected = true;
            firstSelectedCoin = coinCoord;
            [self stopCoinAt: firstSelectedCoin];
        } else {
            ///shown in Listing 4-24
        }
    }

}
```

Listing 4–20 takes the UIGestureRecognizer that triggered this event as an argument. We will use this object to figure out where the user tapped. The first if statement in this task checks that there are remaining turns and that we are currently accepting input. The variable acceptingInput is set to false during animations in order to prevent the user from selecting coins midtransition, which would probably cause the user some confusion. After casting gestureRecognizer to a UITapGestureRecognizer called tapRecognizer, we get the location of the tap by calling locationInView and passing in coinsView. The location of the tap is returned as a CGPoint called location. In order to know which coin was tapped, we convert location to Coord by calling coordFromLocation, shown in Listing 4–21.

Listing 4–21. *CoinsController.m (coordFromLocation:)*

```
-(Coord)coordFromLocation:(CGPoint) location{
    CGRect coinsFrame = [coinsView frame];
    Coord result;
    result.col = location.x / coinsFrame.size.width * [coinsGame colCount];
    result.row = location.y / coinsFrame.size.height * [coinsGame rowCount];

    return result;
}
```

The task coordFromLocation: takes a CGpoint and converts it into a Coord that will tell us which coin the user tapped. To find the coin, the x value of the location is divided by the width of the coinsView and then multiplied by the number of columns. The row is calculated in a similar way.

In Listing 4–20, after we determine which coin was tapped and store the result as the variable coinCoord, we check whether the coin was previously selected. If not, we set isFirstCoinSelected to true, record which coin was selected, and call stopCoinAt:, shown in Listing 4–22.

Listing 4–22. *CoinsController.m (stopCoinAt:)*

```
-(void)stopCoinAt:(Coord)coord{
    UIImageView* coinView = [[coinsView subviews] objectAtIndex:[coinsGame
indexForCoord:coord]];
    NSNumber* coin = [coinsGame coinForCoord:coord];

    UIImage* image = imageForCoin([coin intValue]);
    [coinView stopAnimating];
    [coinView setImage: image];
}
```

The task stopCoinAt: takes the coordinate of the coin to stop spinning. This is done by first finding the appropriate UIImageView representing the coin to be stopped, called coinView. We stop the spinning animation by calling stopAnimation:, but we don't want to just stop it, we want to display the face of the coin. We can find the correct image to use by first calling coinForCoord in coinsGame to figure out the type of coin it is and then calling imageForCoin to return the UIImage we want to use. We simply call setImage on coinView and pass in the UIImage named image. The function imageForCoin simply returns a UIImage based on the int that is passed in, as shown in Listing 4–23.

Listing 4–23. *CoinsController.m (imageForCoin(int))*

```
UIImage* imageForCoin(int coin){
    if (coin == COIN_TRIANGLE){
        return [UIImage imageNamed:@"coin_triangle0001"];
    } else if (coin == COIN_SQUARE){
        return [UIImage imageNamed:@"coin_square0001"];
```

```
    } else if (coin == COIN_CIRCLE){
        return [UIImage imageNamed:@"coin_circle0001"];
    }
    return nil;
}
```

In Listing 4–24, we see the code that is executed when a coin was previously selected.

Listing 4–24. *CoinsController.m (tapGesture:, continued)*

```
if (CoordEqual(firstSelectedCoin, coinCoord)){//re selected the first one.
            isFirstCoinSelected = false;

            [self spinCoinAt:firstSelectedCoin];
        } else {//selected another one, do switch.
            acceptingInput = false;
            [coinsGame setRemaingTurns: [coinsGame remaingTurns] - 1];
            [delegate turnsRemainingDecreased:self with: [coinsGame remaingTurns]];

            isFirstCoinSelected = false;
            secondSelectedCoin = coinCoord;
            [self stopCoinAt:secondSelectedCoin];
            [self doSwitch:firstSelectedCoin With:secondSelectedCoin];
        }
```

If the user selects the same coin, we simply set that coin spinning again and set isFirstCoinSelected to false. However, if the user selects a different coin, it is time to switch the coins. The first thing we do is set acceptingInput to false, so the user doesn't interrupt us as the animations are happening. We also want to decrement the number of remaining turns and inform the delegate that this has happened. We set isFirstCoinSelected to false, because the next time this method is called, there should be no coin selected. After recording the coordinate of the second coin in the variable secondSelectedCoin, we stop the selected coin and call doSwitch:With:, passing in the first and second selected coins. The task doSwitch is shown in in Listing 4–25 in the following section and is responsible for creating the animations of the coins trading places.

This brings us to a discussion of how animations are created without using the static UIView tasks described earlier.

Animating Views with Core Animation

Previously we looked at tasks from the class UIView that helped create animations. These tasks provide an opaque view into how animations are implemented on iOS. This section will take a closer look at how animations are defined with the framework known as Core Animation. Core Animation is the framework responsible for most animations on iOS devices and is the framework used behind the scenes by the static UIView tasks.

The best way to start understanding Core Animation is to continue our example and see how we create the animation of the coins trading places. Figure 4–8 shows the animation of the coins trading places.

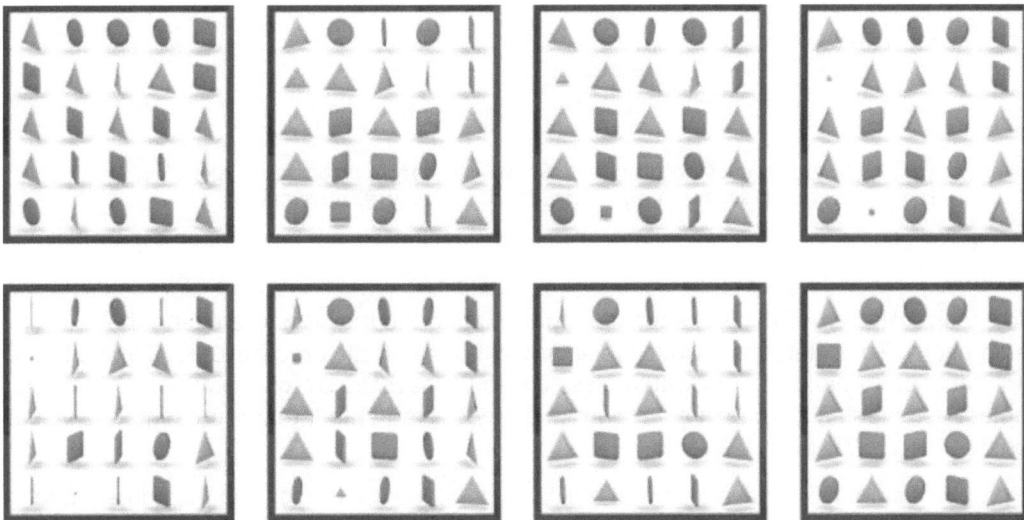

Figure 4–8. *Coins switching locations*

The coin in the first column and second row (a square) is being switched with the coin in the second column and the last row (a triangle). Each coin shrinks until it is too small to see. Then the type of coin switches before growing back to normal size. This is implemented in the task doSwitch:With:, shown in Listing 4–25.

Listing 4–25. *CoinsController.m (doSwitch:With:)*

```
-(void)doSwitch:(Coord)coordA With:(Coord)coordB {
    [coinsGame swap:coordA With:coordB];

    coinViewA = [[coinsView subviews] objectAtIndex:[coinsGame indexForCoord:coordA]];
    coinViewB = [[coinsView subviews] objectAtIndex:[coinsGame indexForCoord:coordB]];


    for (UIView* coinView in [NSArray arrayWithObjects:coinViewA, coinViewB, nil]){

        CABasicAnimation* animScaleDown = [CABasicAnimation
animationWithKeyPath:@"transform.scale"];
        [animScaleDown setValue:@"animScaleDown" forKey:@"name"];
        animScaleDown.fromValue = [NSNumber numberWithFloat:1.0f];
        animScaleDown.toValue = [NSNumber numberWithFloat:0.0f];
        animScaleDown.duration = 1.0;
        animScaleDown.timingFunction = [CAMediaTimingFunction
functionWithName:kCAMediaTimingFunctionEaseIn];


        CABasicAnimation* animScaleUp = [CABasicAnimation
animationWithKeyPath:@"transform.scale"];
        [animScaleUp setValue:@"animScaleUp" forKey:@"name"];
        animScaleUp.fromValue = [NSNumber numberWithFloat:0.0f];
        animScaleUp.toValue = [NSNumber numberWithFloat:1.0f];
        animScaleUp.duration = 1.0;
        animScaleUp.beginTime = CACurrentMediaTime() + 1.0;
        animScaleUp.timingFunction = [CAMediaTimingFunction
functionWithName:kCAMediaTimingFunctionEaseOut];
```

```
        if (coinViewA == coinView){
            [animScaleDown setDelegate:self];
            [animScaleUp setDelegate:self];
        }

        [coinView.layer addAnimation:animScaleDown forKey:@"animScaleDown"];
        [coinView.layer addAnimation:animScaleUp forKey:@"animScaleUp"];

    }
}
```

The task doSwitch:With: is responsible for updating the model and creating the animations. To update the model, we simply call swap:With: on coinsGame. To create the animations, we have to work with a class called CABasicAnimation. A CABasicAnimation object describes a change in a CALayer. A CALayer is an object that represents the visual content of a UIView.

Up until this point, we have said that a UIView provides the content on the screen, and that is still true, but a UIView uses the Core Animation layer to implement how it is drawn. As such, each UIView has a layer property of type CALayer. You need to understand this in order to understand how CABasicAnimation works. Looking at Listing 4–25, you can see that a CABasicAnimation is created by specifying a path. In our case, the path is transform.scale. This is a path into the CALayer object associated with the UIView we wish to animate.

For each UIView, coinViewA and coinViewB, we will be creating two CABasicAnimation objects. When we create the CABasicAnimation object animScaleDown, we specify that this CABasicAnimation will be manipulating the value of scale on the property transform. We specify the starting value of the scale by setting the fromValue of animScaleDown, and we specify the ending value by setting the toValue. The value duration indicates how many seconds we want this animation to take. By specifying the timingFunction, we control the rate at which this animation will take place. In this case, we indicate kCAMediaTimingFunctionEaseIn, which tells the CABasicAnimation to start out slow and then speed up.

The CABasicAnimation animScaleUp is similar to animScaleDown. The big difference is that the fromValue and toValue are opposite. We also set the value for beginTime to 1 second in the future. The idea here is that we want the animation animScaleDown to run for a second, making the coin get small enough to vanish, and then we want the animation animScaleUp to run when animScaleDown is done, scaling the coin back up. The trick will be to swap the images used by the two coin views between these two animations. We can do this by setting self as the delegate to the animation animScaleDown, if we are working with coinViewA, because we need to be notified only once. We also want to know when all of these animations are over, so we also use self as the delegate to animScaleUp. The animations that have had their delegates set will call the task animationDidStop:finished: when they are done, as shown in Listing 4–26.

Listing 4–26. *CoinsController (animationDidStop:finished:)*

```
- (void)animationDidStop:(CAAnimation *)theAnimation finished:(BOOL)flag{
    if ([[theAnimation valueForKey:@"name"] isEqual:@"animScaleDown"]){

        UIImage* imageA = [coinViewA image];
        [coinViewA setImage:[coinViewB image]];
        [coinViewB setImage:imageA];

    } else if ([[theAnimation valueForKey:@"name"] isEqual:@"animScaleUp"]){
        [self checkMatches];
        [self spinCoinAt:firstSelectedCoin];
        [self spinCoinAt:secondSelectedCoin];

    } else if ([[theAnimation valueForKey:@"name"] isEqual:@"animateOffScreen"]){
        [coinsGame randomizeRows: matchingRows];
        [coinsGame randomizeCols: matchingCols];
        [self updateCoinViews];
    }
}
```

Here we can see the task animationDidStop:finished:. Notice that three animations call this task. The first two if statements are for handling the animations described in Listing 4–25. When the animScaleDown animation is done, we swap the UIImages used by UIImageViews to represent the two selected coins.

When the animation animScaleUp is done, we want to check whether there are any matches. Calling checkMatches does this. We also want to start the recently selected coins spinning again. The implementation of checkMatches is shown in Listing 4–27.

Listing 4–27. *CoinsController.m (checkMatches)*

```
-(void)checkMatches{

    matchingRows = [coinsGame findMatchingRows];
    matchingCols = [coinsGame findMatchingCols];

    int rowCount = [coinsGame rowCount];
    int colCount = [coinsGame colCount];

    BOOL isDelegateSet = NO;

    if ([matchingRows count] > 0){

        for (NSNumber* row in matchingRows){
            for (int c=0;c<colCount;c++){
                CABasicAnimation* animateOffScreen = [CABasicAnimation
animationWithKeyPath:@"position.x"];
                [animateOffScreen setValue:@"animateOffScreen" forKey:@"name"];
                animateOffScreen.byValue = [NSNumber
numberWithFloat:coinsView.frame.size.width];
                animateOffScreen.duration = 2.0;
                animateOffScreen.timingFunction = [CAMediaTimingFunction
functionWithName:kCAMediaTimingFunctionEaseIn];

                Coord coord = CoordMake([row intValue], c);
                int index = [coinsGame indexForCoord:coord];
```

```objc
                          UIImageView* coinView = [[coinsView subviews] objectAtIndex: index];

                          if (c == 0){
                              [animateOffScreen setDelegate:self];
                              isDelegateSet = YES;
                          }

                          [coinView.layer addAnimation:animateOffScreen
forKey:@"animateOffScreenX"];

                      }
                  }
              }

          if ([matchingCols count] > 0){

              for (NSNumber* col in matchingCols){
                  for (int r=0;r<rowCount;r++){
                      CABasicAnimation* animateOffScreen = [CABasicAnimation
animationWithKeyPath:@"position.y"];
                      [animateOffScreen setValue:@"animateOffScreen" forKey:@"name"];
                      animateOffScreen.byValue = [NSNumber
numberWithFloat:coinsView.frame.size.height];
                      animateOffScreen.duration = 2.0;
                      animateOffScreen.timingFunction = [CAMediaTimingFunction
functionWithName:kCAMediaTimingFunctionEaseIn];

                      Coord coord = CoordMake(r, [col intValue]);
                      int index = [coinsGame indexForCoord:coord];
                      UIImageView* coinView = [[coinsView subviews] objectAtIndex: index];

                      if (!isDelegateSet && r == 0){
                          [animateOffScreen setDelegate:self];
                      }

                      [coinView.layer addAnimation:animateOffScreen
forKey:@"animateOffScreenY"];
                  }
              }
          }

          int totalMatches = [matchingCols count] + [matchingRows count];
          if (totalMatches > 0){
              [coinsGame setScore:[coinsGame score] + totalMatches];
              [delegate scoreIncreases:self with:[coinsGame score]];
          } else {
              if ([coinsGame remaingTurns] <= 0){
                  //delay calling gameOver on the delegate so the coin's UIImageViews show the
correct coin.
                  [NSTimer scheduledTimerWithTimeInterval:1 target:self
selector:@selector(doEndGame) userInfo:nil repeats:FALSE];
              } else {
                  //all matches are done animating and we have turns left.
                  acceptingInput = YES;
              }
          }
```

We find all matching rows and columns by calling findMatchingRows and findMatchinCols on coinsGame. After we have these arrays of matches, we want to animate all of the coins involved in a match of the screen. If there are matchingRows, we create a new CABasicAnimation for each coin that modifies the position value of x of the underlying CALayer object. Instead of setting the exact starting and ending values of x, we can simply specify the byValue, which we make the width of the coinsView. By specifying the byValue, we can use this animation for all coins in the row, because this animation will simply infer that the fromValue is the starting x value for each coin view, and the toValue with be fromValue + byValue.

After the animations for matching rows and columns are created, we check to see how many matches in total were created on this turn. If this value is bigger than zero, we update the score of the coinsGame object and inform the delegate that the score has changed. If no matches are found, we check whether the game is over. If it is over, we use the NSTimer class to call doEndGame in 1 second. This delay is for aesthetic reasons; it is nice to have a little pause before the high scores are shown. If it is not the end of the game, we simply start accepting input again. Listing 4–28 shows the task doEndGame, which simply informs the delegate that the game is over.

Listing 4–28. *CoinsController.m (doEndGame)*

```
-(void)doEndGame{
    [delegate gameOver:self with: coinsGame];
}
```

The animations created earlier in Listing 4–27 will call the task animationDidStop:finished: when they are done, in the same way the scaling animations did. We use this callback to randomize the matches and call updateCoinViews, shown in Listing 4–29.

Listing 4–29. *CoinsController.m (updateCoinViews)*

```
-(void)updateCoinViews{
    int rowCount = [coinsGame rowCount];
    int colCount = [coinsGame colCount];

    for (NSNumber* row in matchingRows){
        for (int c=0;c<colCount;c++){
            Coord coord = CoordMake([row intValue], c);
            [self spinCoinAt:coord];
        }
    }
    for (NSNumber* col in matchingCols){
        for (int r=0;r<rowCount;r++){
            Coord coord = CoordMake(r, [col intValue]);
            [self spinCoinAt:coord];
        }
    }
    [self checkMatches];
}
```

The task updateCoinView is called when the UIImageViews that represent the coins are fully offscreen. This task goes though each set of rows and columns from that previously matched and sets the coins spinning again in their original locations, but with images that match the new values. Finally, checkMatches is called again, to handle any cases of the new random values creating new matches.

Summary

In this chapter, we explored how to create an input-driven game. This included understanding the management of a game state, the basics of user input, and the use of core animation to create animations. Game state is managed by creating a class to hold the data required to re-create the game in a display-agnostic way. To allow the user to interact with the game, basic gesture recognizers must be added to the correct views. A controller class interprets each user gesture in a game-specific context and produces animations that reflect a change in the game state.

Chapter 5

Quickly Build a Frame-by-Frame Game

We have already looked at a simple game where the action is solely driven by user input. In this chapter, we will be looking at games that animate continuously, regardless of whether the user provides input. Action games are typical examples of such games. We are not going to make a full game in this chapter, but will implement part of one. Figure 5–1 shows one of the scenes we will create.

Figure 5–1. *A frame-by-frame space game*

In Figure 5–1, we see a spaceship dodging some asteroids. You will learn how to animate the different items in the scene, move the ship based on user input, and detect the collisions between the ship and asteroids.

There are many different technologies that have been used over the years to create these types of games. These different technologies concern themselves with how content is drawn on the screen. An example is OpenGL, which provides low-level access to a display. The classes provided by UIKit (UIView and the like) are perfectly suited for these types of games as well. Regardless of which technology is used to update the screen, these types of games generally work in the same way, by creating the loop shown in Figure 5–2.

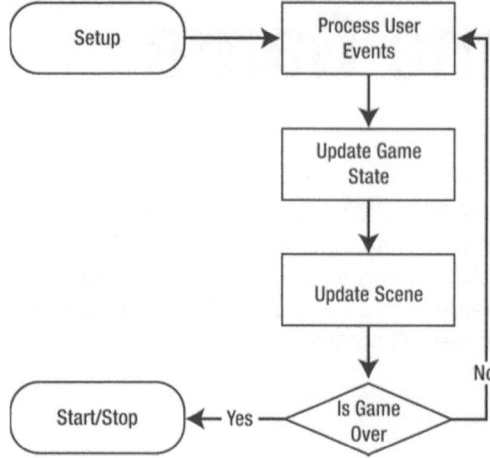

Figure 5–2. *Typical frame-by-frame application loop*

In Figure 5–2, we see that, after the application is set up, a loop is created that is responsible for processing user events, updating the game state, updating the scene on the screen, and, finally, checking if the game is over. When creating an application on iOS that functions this way, we don't have to explicitly create a loop—the application is already running a loop very similar to the one described in Figure 5–1. Let's take a closer look at how we set up an iOS application to accomplish frame-by-frame animations.

Setting Up Your First Frame-by-Frame Animation

The sample code accompanying this chapter is a collection of three examples. Each example builds on the last to illustrate different concepts. When you run the sample code you will see a screen similar to Figure 5–3.

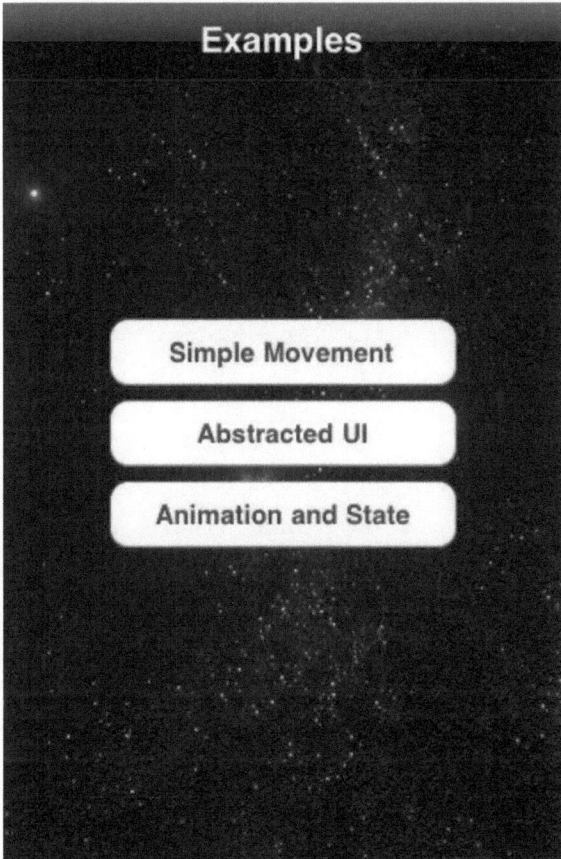

Figure 5–3. *Examples menu*

In Figure 5–3, we see the buttons for each of our three examples. Clicking each button will animate the screen and display the example. Let's start at the top with Simple Movement.

Simple Movement

Simple Movement sets up the application to continuously update the position of a spaceship. The spaceship will move to any point the user taps, as shown in Figure 5–4.

Figure 5–4. *Simple Movement*

In Figure 5–4, we see a spaceship from a top-down perspective with a star field in the background. When the user touches any point on the screen, the ship will animate to that point. This type of animation could easily be accomplished using the type of animations presented in the previous chapter, but this will provide us with a good starting point. In terms of views, what you are seeing in Figure 5–4, ignoring the top navigation bar, is a UIView with two UIImageView subviews. One of the UIImageViews is the background star field and the other is the spaceship. The basic idea here is that we can periodically update the location of the UIImageView of the spaceship to create an animation. Like our other examples, the logic for managing the state of these different UIViews is located in a UIViewController subclass. For this example, the class is called Example01Controller and its header files are shown in Listing 5–1.

Listing 5–1. *Example01Controller.h*

```
#import <UIKit/UIKit.h>
#import <QuartzCore/CADisplayLink.h>
#import "Viper01.h"

@interface Example01Controller : UIViewController {
    CADisplayLink* displayLink;
    Viper01* viper;
}
-(void)updateScene;
-(void)viewTapped:(UIGestureRecognizer*)aGestureRecognizer;
@end
```

We see in Listing 5–1 that the class extends UIViewController and has two fields. The first field is called displayLink and is of type CADisplayLink that is imported from the QuartzCore framework. The second field is called viper and is of type Viper01. It is used to represent the spaceship. There are also two tasks defined: the first is called updateScene and will be used to update the UIViews that make up the scene. The second is called viewTapped, and it is used to manage user input. Let's take a look at the Viper01 header file in Listing 5–2 and see how we are representing the spaceship.

Listing 5–2. *Viper01.h*

```
#import <Foundation/Foundation.h>

@interface Viper01 : UIImageView {
}
@property float speed;
@property CGPoint moveToPoint;
-(void)updateLocation;
@end
```

In Listing 5–2, we see that Viper01 extends UIImageView and has two properties. The speed property describes how fast the viper will move, and the moveToPoint property keeps track of which point on the screen the spaceship is moving. The task updateLocation does the work of incrementally updating the location of the spaceship.

Implementing the Classes

Now that we have seen the definitions of the classes we will be using in this example, let's take a look at the implementation, starting with the viewDidLoad task of Example01Controller class shown in Listing 5–3.

Listing 5–3. *Example01Controller.m (viewDidLoad)*

```
- (void)viewDidLoad
{
    [super viewDidLoad];

    [self setTitle:@"Simple Movement"];

    UITapGestureRecognizer* tapRecognizer = [[UITapGestureRecognizer alloc]
initWithTarget:self action:@selector(viewTapped:)];
    [self.view addGestureRecognizer:tapRecognizer];
```

```
    viper = [Viper01 new];

    CGRect frame = [self.view frame];

    viper.center = CGPointMake(frame.size.width/2.0, frame.size.height/2.0);
    [self.view addSubview:viper];
    [viper setMoveToPoint:viper.center];

    displayLink = [CADisplayLink displayLinkWithTarget:self
selector:@selector(updateScene)];

        [displayLink addToRunLoop:[NSRunLoop currentRunLoop]
forMode:NSDefaultRunLoopMode];
}
```

In Listing 5–3, after setting the title, we create a UITapGestureRecognizer and add it to the root UIView associated with this UIViewController. Adding the UITapGestureRecognizer to the UIView will cause the task viewTapped: to be called when the user taps the screen, giving us a chance to update the location to which the spaceship is moving.

The next step is to create a new instance of Viper01 and place it in the center of the root UIView. We also set the moveToPoint property of viper to the center of the screen, so the viper starts out stationary.

The last thing we have to do is specify a task that we want called periodically so we can update the location of the spaceship and create our animation. We do this by creating a CADisplayLink and specifying the task updateScene. Once we call addToRunLoop, updateScene will be called every time the screen refreshes. In the next section we will look at the class CADisplayLink and NSRunLoop and see what they are all about. Let's look next at the updateScene, as shown in Listing 5–4.

Listing 5–4. *Example01Controller.m (updateScene)*

```
-(void)updateScene{
    [viper updateLocation];
}
```

In Listing 5–4, in the task updateScene, we simply call updateLocation on the object viper. In future examples we will do more in the task updateScene, but for now let's just look at how we implement moving the one ship.

Moving the Spaceship

Each time updateScene is called, we have to update the location of the spaceship. This is accomplished with a little geometry to figure out the new location of the ship for each frame of the animation. Let's just look at updateLocation and understand how that task moves the spaceship. See Listing 5–5.

Listing 5–5. *Viper01.m (updateLocation)*

```
-(void)updateLocation{
    CGPoint c = [self center];

    float dx = (moveToPoint.x - c.x);
    float dy = (moveToPoint.y - c.y);
    float theta = atan(dy/dx);

    float dxf = cos(theta) * self.speed;
    float dyf = sin(theta) * self.speed;

    if (dx < 0){
        dxf *= -1;
        dyf *= -1;
    }

    c.x += dxf;
    c.y += dyf;

    if (abs(moveToPoint.x - c.x) < speed && abs(moveToPoint.y - c.y) < speed){
        c.x = moveToPoint.x;
        c.y = moveToPoint.y;
    }

    [self setCenter:c];
}
```

In Listing 5–5, we see the updateLocation task of the class Viper01. This task is responsible for moving the spaceship one frame's worth of movement toward the point moveToPoint. Because the class Viper01 extends UIImageView, we can move it by simply setting the standard properties used to specify its location relative to its parent. We could use the frame property, but there is a property called center that is a shorthand way of manipulating a UIView's frame.

We start calculating the spaceship's new center location by first getting a copy of its current center and storing that in the variable c. Next, we calculate the difference between the current location and the location we are moving to for both the X and Y axis, storing the values in dx and dy. After looking up the formula in our high school geometry books, we calculate the angle that describes the direction we are moving, by dividing dy by dx and passing the result to the atan (arc tangent) function, storing the result as theta.

We want to move the spaceship speed a number of points in the direction of theta. To calculate what that means in terms X and Y points, we have to do a little geometry. To calculate the number of X points we will be moving, we take the cosine of theta and multiply it by speed, storing the value in dxf. Similarly, to calculate the number of Y points we have to move in this frame, we take the sine of theta and multiple by speed, storing the result in dyf. To finalize these values, we check to see if dx is negative; if it is, we switch the signs of dxf and dyf.

Once dxf and dyf are calculated, we add those values to the x and y parts of the variable c. The last thing we are going to do is set the center property with the new value stored in c. However, if we are less than speed points from the moveToPoint, we will overshoot it. We don't want to overshoot the point, because come next frame we will overshoot again in the opposite direction, causing the spaceship to jitter on the screen. To fix this, we simply set the X and Y values to be identical to moveToPoint's X and Y values if we are close to it. Figure 5–5 shows some of the geometry involved.

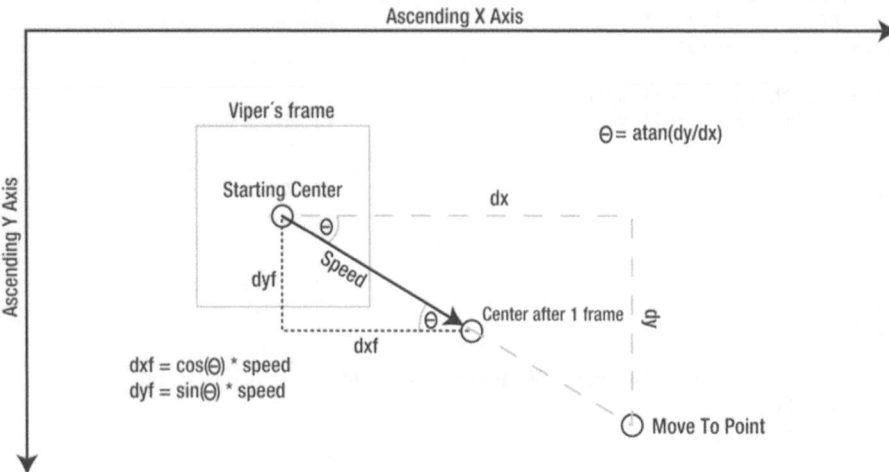

Figure 5–5. *Geometric relations involved in the updateLocation task*

In Figure 5–5, we see the X and the Y axis with the origin on the upper left and the positive Y axis going down. The frame of the spaceship is shown as a grey rectangle with the starting center point in the middle of it. The dashed triangle illustrates how theta is calculated based on the values of dx, dy, and the location of the move to point. The black arrow is of length speed and shows where the center should be located after a single frame. The dotted triangle shows how dxf and dyf are calculated based on the angle.

Responding to a User Tap

We now know we can make the ship move, one frame at a time. But we have not looked at how we tell the ship *where* to move. Recall that we registered a UITapGestureRecognizer on the root UIView of Example01Controlller. This UITapGestureRecognizer is configured to call the task viewTapped:. This is shown in Listing 5–6.

Listing 5–6. *Example01Controller.m (viewTapped:)*

```
-(void)viewTapped:(UIGestureRecognizer*)aGestureRecognizer{
    UITapGestureRecognizer* tapRecognizer = (UITapGestureRecognizer*)aGestureRecognizer;
    CGPoint tapPoint = [tapRecognizer locationInView:self.view];
    [viper setMoveToPoint: tapPoint];
}
```

In Listing 5–6, the first thing we do in viewTapped: is to cast the passed in UIGestureRecognizer to UITapGestureRecognizer. We then find the point where the tap occurred in the root view by calling locationInView:. Once we have the point where the tap occurred, we simply set the moveToPoint property on the viper object. Notice we did not have to interrupt any existing animation—we simply update the state of the viper object and, when its updateLocation task is called again, it will start moving toward this new point. This is a very simple example, but I hope it illustrates an advantage that frame-by-frame animations have over the predefined animations we looked at in the last chapter.

We have looked at how this simple animation is set up and how it is driven by repeated calls to updateScene from Listing 5–4. We should take a little time and understand the classes responsible for making these repeated calls.

Understanding CADisplayLink and NSRunLoop

In general, to create a smooth looking animation, a picture has to be updated at least 25 times a second to avoid the eye seeing each individual frame. That means our game must find a way to call updateScene at 25 times a second to achieve a smooth animation. This could be done by creating an NSTimer that calls updateScene at any rate we choose. If we do that, however, our code that updates the scene will not necessarily be in sync with the hardware's native refresh rate of the screen. The framework Core Animation provides a class specifically designed for creating the type of animation we are trying to create: CADisplayLink.

In Listing 5–3, we created a CADisplayLink, specifying that it should call updateScene whenever the screen is ready to redraw. We then added the CADisplayLink to the same run loop that called viewDidLoad by calling currentRunLoop on the class NSRunLoop. The class NSRunLoop manages a thread to produce the kind of loop shown in Figure 5–1. An NSRunLoop is responsible for processing input from the user and from the CADisplayLink and scheduling when corresponding task in Example01Controller should be called. We don't have to understand too much about how NSRunLoop works; we just have to understand that the tasks viewTapped: and updateScene will be called by the same thread so we don't have to worry about multi threaded complexity.

As mentioned, CADisplayLink—working with our main threads NSRunLoop—causes the task updateScene to be called once for each time the screen redraws. The property duration of CADisplayLink reports, in seconds, the time between each screen redraw. A little investigation shows that this property reports a slightly different value each time updateScene is called, but hovers around the value 0.0166648757. This value is just shy of 60 frames per second, so we are well above the required 25 frames per second.

In more complex applications, it is entirely likely that the state of the game cannot be updated In 1.5 milliseconds; there may simply be too much book keeping. If this is the case, you can set the frameInterval property of CADisplayLink to value greater than one. This will cause CADisplayLink to skip screen redraws. Figure 5–6 shows four different scenarios that can occur when using the CADisplayLink.

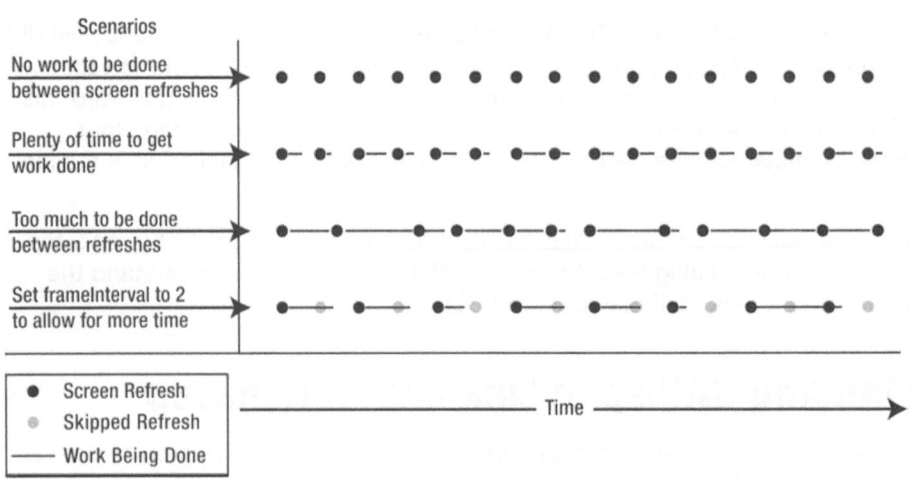

Figure 5–6. *Screen refresh scenarios*

In Figure 5–6, we see four scenarios listed on the left. To the right of each scenario are a number of dots representing screen refreshes for that scenario. Looking at the top scenario, where there is no work to be done between screen refreshes, each dot is equally spaced, meaning that the screen refreshes happen at even intervals as time progresses to the right. The second scenario shows the case when there is a little bit of work to do between screen refreshes, indicated by the lines coming off to the right of each dot. The length of each line is shorter than the distance between each dot, indicating that the work is done in the time between screen refreshes. The third scenario shows the case where there is too much work to get done between screen refreshes, but the programmer did not account for this. Each time the work exceeds the normal amount of time between screen refreshes, the drawing of the next screen is delayed. Because the amount of time it takes to update a game between frames varies, the rate at which the screen is refreshed becomes irregular. When the screen refresh is irregular, the animation looks choppy and is perceived poorly by the user. To avoid this, we configure `CADisplayLink` to only trigger an update every other frame by setting `frameInterval` to 2. This doubles the amount of time we have to get our work done, allowing for a smooth animation.

This strategy can seem counter-intuitive to some developers, because in the fourth scenario there can be long stretches of time where no work is being done. In fact, for a given stretch of time, scenario 3 gets more frames rendered than scenario 4. The fact of the matter is that the choppy updates are almost always more annoying to a user than a game that does not move as fast. Also keep in mind that the native refresh rate for iOS devices is 60 frames a second, so if every other frame is skipped, the animation rate is still 30 frames per second. That's plenty to create a convincing animation.

We have looked at a simple frame-by-frame animation, and you should now understand the basic principles. The next two examples will build on this pattern to create animations with more than one thing on the screen, as well as do a little refactoring to start making a framework suitable for complex games.

Abstracting the UI

In the previous example, we looked at the basics of creating a frame-by-frame animation. We looked how we can use the class CADisplayLink in conjunction with the existing event loop to call a task where we can implement our animation. We animated a single image around the screen based on user input. This was a very simple example. I left out a number of practical considerations so we could focus on the basic idea of creating animations in this way. In this section, we will do a little refactoring in order to better support more complicated games. Figure 5–7 shows the example we will be building.

Figure 5–7. *Abstracted UI*

In Figure 5–7, we see some differences from the first example. The first thing to note is that the play area of the game does not take up the full screen. The slider at the bottom of the screen changes the size of the play area without changing how the game plays. There is also a second item on the screen, an asteroid. Asteroids are created every 10 seconds and start at the top of the screen, moving straight down. If an asteroid collides with the spaceship, the spaceship resets to the center of the screen and the asteroid is removed. The object of this simple game is to avoid the incoming asteroids.

One of the goals of this section is to explain how the game logic can be separated from how the game is drawn on the screen. In the first example we simply extended UIImageView and added our logic about the spaceship. This worked well enough, but there are lots a good reasons for separating the logic of the game from the how it is displayed. For starters, if you want to run this game on an iPad instead of an iPhone, you probably want to make the play area bigger. Also, to support things like zooming into the game or showing a mini map, it makes sense to have the location of each item in the game stored in a display-agnostic way. Down the road, you may decide you want to port your game to another platform. Having a clean separation of game logic from display logic will save you a lot of work in such a case.

To support this abstraction between the game logic and the display logic we introduce a new class called Actor02. Normally this class would simply be called Actor. The 02 in the name simply indicates it is part of the second example in this chapter. The following section explains what an actor is and how it helps implement an abstraction between the game logic and display.

Understanding Actors

Actors are the moving items in a game. In our case, the spaceship and the asteroid are actors. This is distinguished from other things that might be in a game, like a label for displaying the score, or other components.

Actors are sometimes called sprites, but I prefer the term actor, because the word sprite can also describe the image used to display the game item. In causal conversation, it makes little difference what term you use—but let's call them actors.

Example2Controller Overview

In our example, we will define the class Actor02 and show how we can subclass Actor02 to create our game items, the spaceship and the asteroids. Listing 5–7 shows the header file for the class Actor02.

Listing 5–7. *Actor02.h*

```
#import <Foundation/Foundation.h>

@class Example02Controller;
long nextId;
@interface Actor02 : NSObject {

}
@property (nonatomic, retain) NSNumber* actorId;
@property (nonatomic) CGPoint center;
@property (nonatomic) float speed;
@property (nonatomic) float radius;
@property (nonatomic, retain) NSString* imageName;
```

```
-(id)initAt:(CGPoint)aPoint WithRadius:(float)aRadius AndImage:(NSString*)anImageName;
-(void)step:(Example02Controller*)controller;
-(BOOL)overlapsWith: (Actor02*) actor;

@end
```

In Listing 5–7, we see the header file for the class Actor02. There are five properties that represent the basic attributes required for this example. The property actorId is used to identify the actor; we will see how we use this to map each Actor02 to an UIImageView. The property center tells us the location of each Actor02. The property speed indicates how fast this actor moves. The property radius represents the size of this actor. Finally, the property imageName describes how this actor should be drawn on the screen.

The type of the property center is CGPoint. Strictly speaking, we will not be directly using this CGPoint to set the location of a UIView or other Core Graphics component. But CGPoint is a simple way to wrap an X and Y value, and we are already familiar with it. For this example, it makes little difference if we use CGPoint or a struct of our creation to represent a point. However, if you wanted to make sure your code was not dependent on Core Graphics in anyway, you would want to define your own type for representing a point.

Also of note in Listing 5–7 is that an Actor02 does not have a width or height, just a radius. This is done to simplify the example, and also because many games only require a single dimension to describe the size of an actor. If you build your own game framework, you may choose to have your actors include a width and height—it really depends on the game. Certainly if you look at prebuilt game engines you will find that their base type as all sorts of features you may or may not use. It may make a lot of sense to use a game engine, but here we are trading complexity for understanding.

A Simple Actor

Listing 5–7 shows that the Actor02 has three tasks define. Listing 5–8 shows the implementation of these tasks.

Listing 5–8. *Actor02.m*

```
#import "Actor02.h"

@implementation Actor02
@synthesize actorId;
@synthesize center;
@synthesize speed;
@synthesize radius;
@synthesize imageName;

-(id)initAt:(CGPoint)aPoint WithRadius:(float)aRadius AndImage:(NSString*)anImageName{
    self = [super init];
    if (self != nil){
        [self setActorId:[NSNumber numberWithLong:nextId++]];
        [self setCenter:aPoint];
        [self setRadius:aRadius];
        [self setImageName:anImageName];
    }
```

```
        return self;
}
-(void)step:(Example02Controller*)controller{
    //implemented by subclasses.
}
-(BOOL)overlapsWith: (Actor02*) actor {
        float xdist = abs(self.center.x - actor.center.x);
        float ydist = abs(self.center.y - actor.center.y);
    float distance = sqrtf(xdist*xdist+ydist*ydist);
    return distance < self.radius + actor.radius;
}
@end
```

In Listing 5–8 we see the implementation of the class Actor02. The task initAt:WithRadius:AndImage: is used to initialize an Actor02 with some basic information. This task also assigns a random actorId value by using the value nextId and incrementing it. Because nextId is of type long, you will have to create an incredible number of actors before you get two actors with the same id.

The task step: takes an instance of Example02Controller and is used incrementally update the location and state of this Actor02. Subclasses will provide an implementation to this task to provide custom behavior and animations.

The last task, overlapsWith:, is used to check to see if this Actor02 overlaps another Actor02. This task is used to check for collisions in this example.

Actor Subclasses: Viper02

As we know, there are two types of actors in this example, the asteroids and the spaceship. Let's take a look at the class Viper02 and see how it is different from Viper01. The header for Viper02 is shown in Listing 5–9.

Listing 5–9. *Viper02.h*

```
@class Example02Controller;
@interface Viper02 : Actor02 {

}
@property CGPoint moveToPoint;

+(id)viper:(Example02Controller*)controller;
-(void)doCollision:(Actor02*)actor In:(Example02Controller*)controller;
@end
```

In Listing 5–9, we see that the header for Viper02 is not much different from Viper01, as shown in Listing 5–2. Viper02 extends the class Actor02 instead of UIImageView. We have the property moveToPoint, which is the same, but the property speed is absent, because that is inherited from Actor02. We have a new constructor to simplify the creation of a Viper02 and a new task called doCollision:In: that is used to handle the new functionality of colliding with an asteroid. Listing 5–10 shows the implementation of the class Viper02.

Listing 5–10. *Viper02.m*

```
#import "Viper02.h"
#import "Example02Controller.h"


@implementation Viper02
@synthesize moveToPoint;

+(id)viper:(Example02Controller*)controller{

    CGSize gameAreaSize = [controller gameAreaSize];
    CGPoint center = CGPointMake(gameAreaSize.width/2, gameAreaSize.height/2);

    Viper02* viper = [[Viper02 alloc] initAt:center WithRadius:16 AndImage:@"viper"];
    [viper setMoveToPoint:center];
    [viper setSpeed:.8];

    return [viper autorelease];
}

-(void)step:(Example02Controller*)controller{
    CGPoint c = [self center];

    float dx = (moveToPoint.x - c.x);
    float dy = (moveToPoint.y - c.y);
    float theta = atan(dy/dx);

    float dxf = cos(theta) * self.speed;
    float dyf = sin(theta) * self.speed;

    if (dx < 0){
        dxf *= -1;
        dyf *= -1;
    }

    c.x += dxf;
    c.y += dyf;

    if (abs(moveToPoint.x - c.x) < self.speed && abs(moveToPoint.y - c.y) < self.speed){
        c.x = moveToPoint.x;
        c.y = moveToPoint.y;
    }

    [self setCenter:c];
}

-(void)doCollision:(Actor02*)actor In:(Example02Controller*)controller{
    CGSize gameAreaSize = [controller gameAreaSize];
    CGPoint centerOfGame = CGPointMake(gameAreaSize.width/2, gameAreaSize.height/2);
    self.center = centerOfGame;
    self.moveToPoint = centerOfGame;

    [controller removeActor:actor];
}
@end
```

In Listing 5–10, we see that the constructor task `viper:` creates a new `Viper02` object and uses the super task `initAt:WithRadius:AndImage` to set up the basic properties of the object. Also, the property `moveToPoint` is set to be its current location and the speed is set to .8.

The task `step:` is the task that is called once per frame of the animation. The implementation of this task is identical to the implementation of the task `updateLocation:` from the class `Actor01`, as shown in Listing 5–5.

The last task in Listing 5–10 is the task `doCollision:In:`, which is called when this `Viper02` is overlapping with an `Asteroid02`. This task resets the `Viper02` back to the starting location at the center of the game area and removes the offending `Asteroid02` from the game.

Actor Subclass: Asteroid02

After looking at the class `Viper02`, it may not be clear exactly how this is different from the `Viper01`. Sure, some things have moved around, but it looks very similar: it is still basically responsible for updating a center point. After we look at the class `Asteroid02`, we will see how these actor classes are used to describe a scene, versus being the scene itself. The header to the class `Asteroid02` is shown in Listing 5–11.

Listing 5–11. *Asteroid02.h*

```
#import <Foundation/Foundation.h>
#import "Actor02.h"

NSMutableArray* imageNameVariations;

@interface Asteroid02 : Actor02 {

}
+(NSMutableArray*)imageNameVariations;
+(id)asteroid:(Example02Controller*)controller;
@end
```

In Listing 5–11, the header file for the class `Asteroid02` shows that it extends `Actor02` and has two static tasks. There is also an array called `imageNameVariations`. The array, in combination with the task `imageNameVariations`, allows us to create `Asteroid02` with similar, but different graphics. Let's take a look at the implementation of this class and see what that means, as shown in Listing 5–12.

Listing 5–12. *Asteroid02.m*

```
#import "Asteroid02.h"
#import "Example02Controller.h"

@implementation Asteroid02

+(id)asteroid:(Example02Controller*)controller{

    CGSize gameAreaSize = [controller gameAreaSize];

    float radius = arc4random()%8+8;
```

```objc
    float x = radius + arc4random()%(int)(gameAreaSize.width+radius*2);
    CGPoint center = CGPointMake(x, -radius);
    NSString* imageName = [[Asteroid02 imageNameVariations]
objectAtIndex:arc4random()%3];


    Asteroid02* asteroid = [[Asteroid02 alloc] initAt:center WithRadius:radius AndImage:
imageName];

    float speed = (arc4random()%10)/10.0 + .1;

    [asteroid setSpeed: speed];

    return asteroid;

}
+(NSMutableArray*)imageNameVariations{
    if (imageNameVariations == nil){
        imageNameVariations = [NSMutableArray new];
        [imageNameVariations addObject:@"AsteroidA"];
        [imageNameVariations addObject:@"AsteroidB"];
        [imageNameVariations addObject:@"AsteroidC"];
    }
    return imageNameVariations;
}

-(void)step:(Example02Controller*)controller{
    CGPoint newCenter = self.center;
    newCenter.y += self.speed;
    self.center = newCenter;


    if (newCenter.y - self.radius > controller.gameAreaSize.height){
        [controller removeActor: self];
    }
}
@end
```

In Listing 5–12, the constructor task asteroid: creates a new Asteroid02 object and sets up its starting state. For variety, each asteroid is assigned a random radius and a random X location. The starting Y location is set to negative radius, so the asteroid starts just off the top of the game area. Each Asteroid02 is also assigned a random speed.

The task step: implements the motion of the Asteroid02—it simply increases the Y value of the property center by speed until it reaches the bottom of the game area, where it removes itself from the game.

As mentioned, we want the graphic used to represent the Asteroid02 to have some variety as well. To implement this, we have the task imageNameVariations that lazily populates the global NSMutableArray imageNameVariations. In the task asteroid:, a random string is pulled from this array and set as the Asteroids. Figure 5–8 shows the image files used in this example.

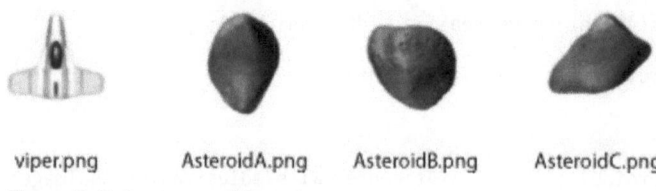

viper.png   AsteroidA.png   AsteroidB.png   AsteroidC.png

Figure 5–8. *Images used for actors*

In Figure 5–8, we see the images used for the actors in this example. Each of these images is specified by the property imageName of class Actor02. In the case of the asteroids, we provide three different images that are assigned randomly to each new Asteroid02 object.

We have taken a look at the actor class and see how they store the basic information for each item in the game. We now want to take look these classes are drawn on the screen.

Drawing Actors on the Screen

We now have a basic presentation of the actors involved in this example. The class that binds them together to create our game is Example02Controller. Like other UIViewController classes, Example02Controller is responsible for displaying data on the screen—in our case, the different actors. The trick in Example02Controller will be to create and update a number of UIImageViews, one for each actor. Figure 5–9 shows how the iPhone version of this class is set up in Interface Builder.

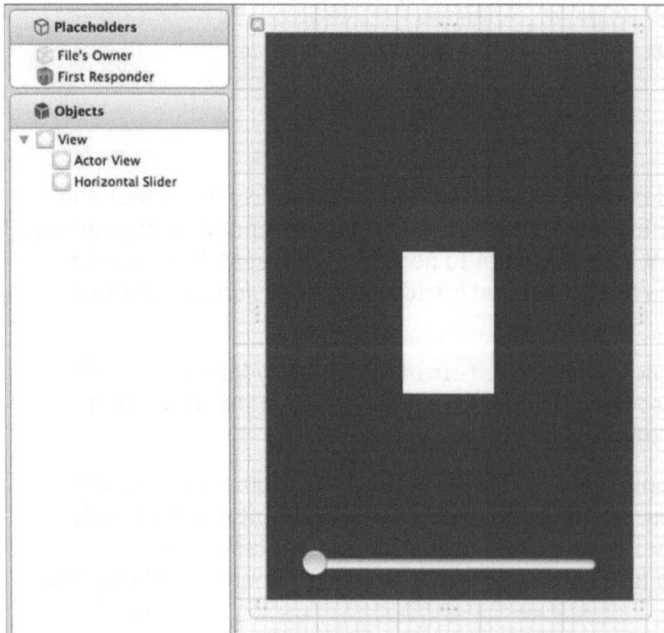

Figure 5–9. *Example02Controller_iPhone.xin in Interface Builder*

In Figure 5–9, we see on the left that the root view for this UIViewController has two subviews. The first is called Actor View and is the white region in the middle of the grey region on the right. This is the view where the game will take place—all actors will be represented as UIImageViews that are subviews of Actor View. The other component is the UISlider at the bottom of the grey area. This is used to control the size of the Actor View, allowing us to demonstrate that we have indeed abstracted the representation of the game, from the game itself. The header to the class Example02Controller will give us an overview of how this is achieved, and is shown in Listing 5–12.

Listing 5–13. *Example02Controller.h*

```
#import <UIKit/UIKit.h>
#import <QuartzCore/CADisplayLink.h>
#import "Viper02.h"
#import "Actor02.h"
#import "Asteroid02.h"

@interface Example02Controller : UIViewController {
    IBOutlet UIView *actorView;

    CADisplayLink* displayLink;

    //Managing Actors
    NSMutableArray* actors;
    NSMutableDictionary* actorViews;

    //Game Logic
    Viper02* viper;
    long stepNumber;

}
@property (nonatomic) CGSize gameAreaSize;


-(void)updateScene;
-(void)removeActor:(Actor02*)actor;
-(void)addActor:(Actor02*)actor;
-(void)updateActorView:(Actor02*)actor;
- (void)tapGesture:(UIGestureRecognizer *)gestureRecognizer;
- (IBAction)sliderValueChanged:(id)sender;

@end
```

In Listing 5–13, we see the UIView actorView and that it is an IBoutlet wired up from the XIB file shown in Figure 5–9. At the bottom of the file, we see the task sliderValueChanged:, which is called when the slider from Figure 5–9 is moved. We also see the familiar displayLink object of type CADisplayLink that will be used to configure redrawing the screen. The NSMutableArray actors is used to store all current actors in the game, which are added and removed with the tasks addActor: and removeActor:. The NSMutableDictionary actorViews is used to map each actor to each UIImageView that represents it. The variable viper of type Viper02 is a references to our spaceship, and the last variable, stepNumber, keeps track of how many frames of the game we have stepped through. Like most UIViewController classes, understanding how it works starts by looking at the viewDidLoad task shown in Listing 5–14.

Listing 5–14. *Example02Controller.m (viewDidLoad)*

```
- (void)viewDidLoad
{
    [super viewDidLoad];
    [self setGameAreaSize:CGSizeMake(160, 240)];
    actors = [NSMutableArray new];
    actorViews = [NSMutableDictionary new];

    Actor02* background = [[Actor02 alloc] initAt:CGPointMake(80, 120) WithRadius:120
AndImage:@"star_field_iphone"];
    [self addActor: background];

    viper = [Viper02 viper:self];
    // [viper setMoveToPoint:viper.center];
    [self addActor:viper];

    stepNumber = 0;

    UITapGestureRecognizer* tapRecognizer = [[UITapGestureRecognizer alloc]
initWithTarget:self action:@selector(tapGesture:)];
    [tapRecognizer setNumberOfTapsRequired:1];
    [tapRecognizer setNumberOfTouchesRequired:1];

    [actorView addGestureRecognizer:tapRecognizer];

    displayLink = [CADisplayLink displayLinkWithTarget:self
selector:@selector(updateScene)];
        [displayLink addToRunLoop:[NSRunLoop currentRunLoop]
forMode:NSDefaultRunLoopMode];

}
```

In Listing 5–14, after calling the super implementation of viewDidLoad, we set the property gameAreaSize. In the first example, we used the coordinate space of a UIView to describe the location of our spaceship. In this example, we have to specify the size of play area independent of any UIView. Note, however, that the size 160 x 240 is exactly one quarter of the number of points of an iPhone. Even though we want to separate the coordinate system of your game from the coordinate system of the UIView hierarchy, the fact is that an iPhone has a screen ration of 2:3, and any game we want to play on that device should consider that.

After specifying the gameAreaSize, we initialize actors and actorViews. Then we add our first actor. To add the star field behind the other actors, we can add new Actor02 to the scene, specifying the star_field_iphone image. Because the base Actor02 class does not specify any behavior, this actor will simply sit in the background, providing the nice stars. The advantage to using an actor for the background is that it will scale and stretch along with the other actors. The disadvantage is that it will be treated like the other actors in the game and consume a little computation time.

After adding the background, we create our Viper02, set it the variable viper, and add it to the scene. We store a reference to it so we can access it easily when the user touches the screen. This is an operation we know we are going to do a lot, so there is no point searching though all of the actors each time to find the correct one. For any given game, some thought will have to be put into which actors are references in special ways. There is a balance between ease of programming, speed of execution, and memory usage that must be considered. Unfortunately, there are no hard and fast rules about what additional data structures should be created. I try and to do as little optimization as possible on my first pass through a game (or application) and wait until the game is more complete to try and figure out the hot spots in an application. Optimizations early in the development process can complicate feature development; however, optimization at the end of development can be a lot more complicated, and might never get done. The two steps in Listing 5–14 are to add a UITapGestureRecognizer to the actorView and to set up the CADisplayLink. These steps are done in exactly the same way as they were done in the first example.

Before we look at how the scene is updated, let's take a look at the tasks addActor: and removeActor: so we can complete our understanding of how actors are added and removed from the scene. Listing 5–15 shows these two tasks.

Listing 5–15. *Example02Controller.m (addActor: and removeActor:)*

```
-(void)addActor:(Actor02*)actor{
    [actors addObject:actor];
}

-(void)removeActor:(Actor02*)actor{
    UIImageView* imageView = [actorViews objectForKey:[actor actorId]];
    [actorViews removeObjectForKey:actor];

    [imageView removeFromSuperview];
    [imageView release];

    [actors removeObject:actor];

    [actor release];
}
```

In Listing 5–15, the task addActor simply adds the actor to the NSMutableArray actors. If you wanted to keep track of actors as they were added, this is where you would do that. The task removeActor finds the UIImageView used to draw it on the screen and removes it from the UIView actorViews. The UIImageView is also removed from the scene by calling removeFromSuperview, and is finally released. The object actor is also removed from the NSMutableArray actors and is released. Now that we see how actors are added and removed from the game, let's take a look at the task updateScene that is called periodically to advance our game a single frame, as shown in Listing 5–16.

Listing 5–16. *Example02Controller.m (updateScene)*

```
-(void)updateScene{
    if (stepNumber % (60*10) == 0){
        [self addActor:[Asteroid02 asteroid:self]];
    }

    for (Actor02* actor in actors){
        [actor step:self];
    }
```

```
    for (Actor02* actor in actors){
        if ([actor isKindOfClass:[Asteroid02 class]]){
            if ([viper overlapsWith:actor]){
                [viper doCollision:actor In:self];
                break;
            }
        }
    }

    for (Actor02* actor in actors){
        [self updateActorView:actor];
    }
    stepNumber++;
}
```

In Listing 5–16, we see the task updateScene, this task is the hart beat of the game, it gets called about 60 times a second and is where we advance the game along. The first thing we do is check to see if we want to add a new asteroid to the game. This is done by taking stepNumber and moding it by 600 and seeing if the value is zero. In effect, this adds a new Asteroid02 to the scene every 10 seconds, since the game is running at about 60 frames a second.

After testing if a new asteroid should be added, we iterate through all of the actors in the game and call step: on them. This gives each actor a change to advance its state based on its particular behavior. The background will do nothing, the asteroids will move downward, and the spaceship will move toward its moveToPoint. After updating the location of each actor we want to test to see if there was a collision. This is done by again iterating through all of the actors, finding the ones that are asteroids, and checking for a collision state. For this simple example, iterating over all of the actors to find a collision is ok. However, in a more complex application it is probably important to improve this algorithm (such as it is). Improvements can include keeping track of all asteroids in their own array and possible sorting them by location.

Updating the UIView for Each Actor

We have tested if there were any collisions and now we have to figure out the new location of each actor's UIImageView. This is done in the updateActorView: task, as shown in Listing 5–17.

Listing 5–17. *Example02Controller.m (updateActorView:)*

```
-(void)updateActorView:(Actor02*)actor{
    UIImageView* imageView = [actorViews objectForKey:[actor actorId]];

    if (imageView == nil){
        UIImageView* imageView = [[UIImageView alloc] initWithImage:[UIImage
imageNamed:[actor imageName]]];
        [actorViews setObject:imageView forKey:[actor actorId]];
        [imageView setFrame:CGRectMake(0, 0, 0, 0)];
        [actorView addSubview:imageView];
    }

    float xFactor = actorView.frame.size.width/self.gameAreaSize.width;
```

```
    float yFactor = actorView.frame.size.height/self.gameAreaSize.height;

    float x = (actor.center.x-actor.radius)*xFactor;
    float y = (actor.center.y-actor.radius)*yFactor;
    float width = actor.radius*xFactor*2;
    float height = actor.radius*yFactor*2;
    CGRect frame = CGRectMake(x, y, width, height);
    [imageView setFrame:frame];

}
```

In Listing 5–17, the first thing we do is find which UIImageView is representing the actor that was passed in. We find it by looking up the UIImageView in the NSMutableArray actorViews, using the actor's actorId as the key. If we don't find a UIImageView, then this actor must have just been added, so we have to create it.

Creating the UIImageView is as simple as creating a UIImage based on the imageName of the actor and creating a new UIImageView with it. We then put the UIImageView into the NSMutableDictionary actorViews, again using the actor's ID as the key. Lastly, we add the new UIImageView to actorView. We set the frame of the new UIImageView to a size zero frame to prevent some redraw issues. The frame will be updated shortly anyway.

Placing UIImageView in the Screen

After creating the UIImageView (if required) we have to figure out where on the screen it should be drawn. The first step is to figure out the relationship between the size of the game area and the actual size of actorView on the screen. We can simply divide width of the actorView by the width of the game area to establish our xFactor. We do the same operation with the heights to figure out our yFactor. Once we have these ratios we can calculate the frame of the UIImageView. Once the four values that make up the new frame of the UIImageView we set it. Figure 5–10 illustrates how these values are calculated.

In Figure 5–10, on the left we see the game area and on the right is the CGRect that describes the actorView's frame. The circle on the left is an Actor02 that needs to be converted into CGRect on actorView. We start by figuring out the origin of the GCFrame by first finding the upper-left point of the Actor02. We find the X value of the upper-left point by subtracting the radius of the Actor02 from the Actor02's center X value. To find the Y value, we subtract the radius from the Actor02's center Y values. To convert these points into the coordinate space of actorView, we simply multiple by the X value of the upper left point by xFactor and multiple the Y value by yFactor. The values xFactor and yFactor are the ratios between the gameAreaSizes width and height and the actorView's frame's width and height. To find the size of actorView's frame, we simply multiply the radius by xFactor and yFactor to find the width and height.

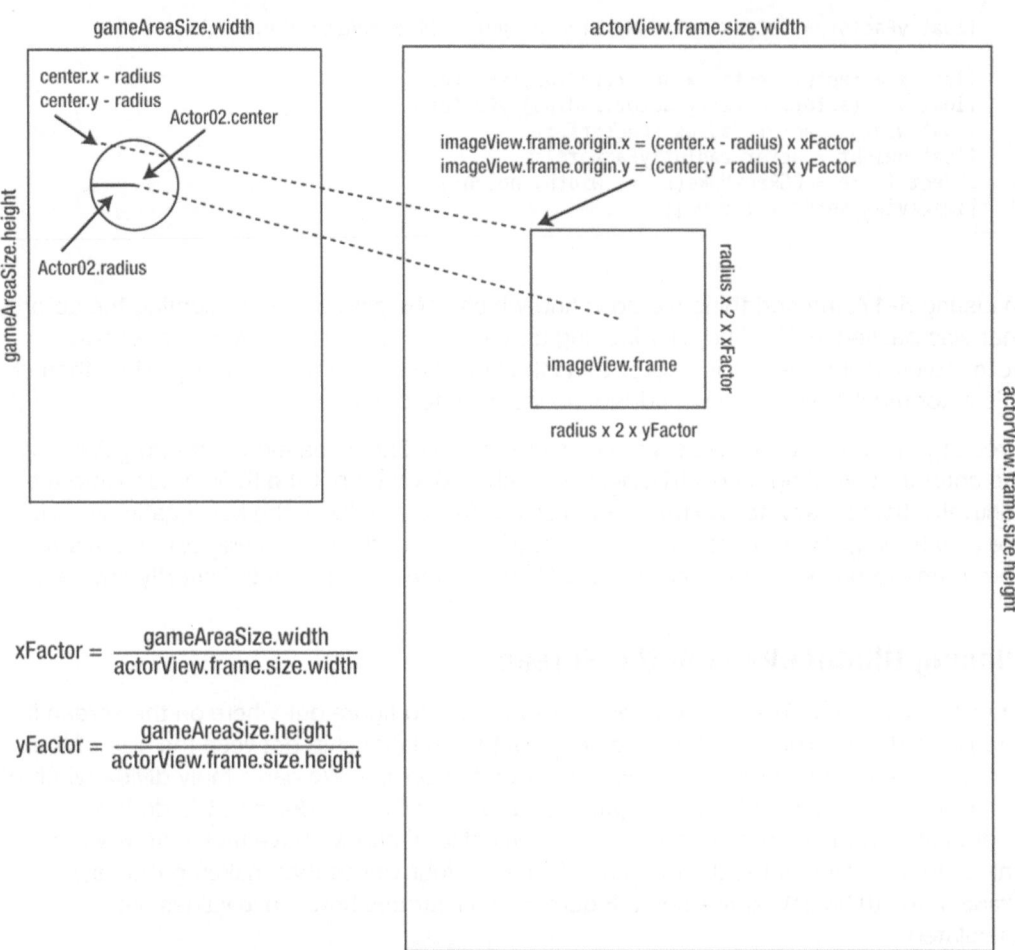

Figure 5–10. *Converting game coordinates into screen coordinates*

When the user touches the UIView actorView, we have to do this process in reverse: convert the point on the actorView to a point with our game space. This conversion happens in the task tapGesture:, as shown in Listing 5–18.

Listing 5–18. *Example02Controller.h (tapGuesture:)*

```
- (void)tapGesture:(UIGestureRecognizer *)gestureRecognizer{
    UITapGestureRecognizer* tapRecognizer = (UITapGestureRecognizer*)gestureRecognizer;

    CGPoint pointOnView = [tapRecognizer locationInView:actorView];

    float xFactor = actorView.frame.size.width/self.gameAreaSize.width;
    float yFactor = actorView.frame.size.height/self.gameAreaSize.height;

    CGPoint pointInGame = CGPointMake(pointOnView.x/xFactor, pointOnView.y/yFactor);

    [viper setMoveToPoint:pointInGame];
}
```

In Listing 5–18, we find where the user touched by calling locationInView: on tapRecognizer and passing in actorView and storing the result in pointOnView. After recalculating xFactor and yFactor values, we simply divide pointOnView's X and Y values by xFactor and yFactor, giving us pointInGame. Using this value we simply, set the moveToPoint property of viper with pointInGame.

To wrap up this example, let's take a look at the code that changes the size of actorView, as shown in Listing 5–19.

Listing 5–19. *Example02Controller.m (sliderValueChanged:)*

```
- (IBAction)sliderValueChanged:(id)sender {
    UISlider* slider = (UISlider*)sender;
    float newWidth = [slider value];
    float newHeight = gameAreaSize.height/gameAreaSize.width*newWidth;

    CGRect parentFrame = [[actorView superview] frame];
    float newX = (parentFrame.size.width-newWidth)/2.0;
    float newY = (parentFrame.size.height-newHeight)/2.0;

    CGRect newFrame = CGRectMake(newX, newY, newWidth, newHeight);
    [actorView setFrame:newFrame];
}
```

The task sliderValueChanged: in Listing 5–19 is called when the user adjust the slider at the bottom of the screen. The slider is configured to have a value between 80 and 320, which we use as the value as newWidth. We calculate the newHeight based on the newWidth, preserving our aspect ratio. Once we have the new width and height for actorView, we find the X and Y values that will keep actorView centered. Once we have all of the value values for the actorView's new frame, we simply set it.

Actor State and Animations

Now that we have a basic set of animations, it is time to add a little life to actors. We are going to expand on the last examples and add two effects. The first is to make the asteroids look like they are tumbling through space. The second is to make the spaceship rotate before moving. Both of these techniques will update the image used to represent each actor to create the visual effect. We will also add some state logic to the spaceship to we can keep track of what the ship should be dong, staying put, rotating, or traveling to the target point. Figure 5–10 shows this next example in action.

Figure 5-11. *Adding animations and state to actors*

The Tumbling Effect

In Figure 5-11 we see the spaceship thrusting toward some point at the upper left. There are four asteroids in the scene, each with a different graphic. When you run the example, you will see that they appear to be tumbling. This tumbling effect is achieved by changing the image used to represent the asteroid every few frames of the animation. The images used create an animation, like a flipbook. Figure 5-12 shows the images used to create the animation.

Figure 5–12. *Images that make up the animation for asteroid variant B*

In Figure 5–12, we see 31 images of asteroids. If the images are showed in succession, moving from the top left to the bottom right, it will appear as if the asteroid is rolling. The animation is a loop, so the first image can be shown after the last, creating a smooth animation. To understand how we implement this animation, let's take a look at the header file for the update class Actor03, as shown in Listing 5–20.

Listing 5–20. *Actor03.h*

```
#import <Foundation/Foundation.h>

@class Example03Controller;
long nextId;
@interface Actor03 : NSObject {

}
@property (nonatomic, retain) NSNumber* actorId;
@property (nonatomic) CGPoint center;
@property float rotation;
@property (nonatomic) float speed;
@property (nonatomic) float radius;
@property (nonatomic, retain) NSString* imageName;
@property (nonatomic) BOOL needsImageUpdated;

-(id)initAt:(CGPoint)aPoint WithRadius:(float)aRadius AndImage:(NSString*)anImageName;
-(void)step:(Example03Controller*)controller;
-(BOOL)overlapsWith: (Actor03*) actor;

@end
```

In Listing 5–20, we see a number of small changes. We have added two new properties: rotation and needsImageUpdate. The rotation property does not refer to the rotation of the asteroid; the rotation property will be used by the updated Viper03 class to indicate what direction it is facing. The property needsImageUpdated is a property that indicates that a new UIImage used should be used for this actor. By setting this value to Yes, we cause each asteroid to ask for a new image while it is being animated. Let's take a look at the changes in the `Asteroid03` class that take advantage of this property—see Listing 5–21.

Listing 5–21. *Asteroid03.h and Asteroid03.m (partial)*

```
//From Asteroid.h
#define NUMBER_OF_IMAGES 31
…

@property (nonatomic) int imageNumber;
@property (nonatomic, retain) NSString* imageVariant;
//From Asteroid.m
-(NSString*)imageName{
    return [[imageVariant stringByAppendingString:@"_"]
stringByAppendingString:[NSString stringWithFormat:@"%04d", self.imageNumber]];
}

-(void)step:(Example03Controller*)controller{
    if ([controller stepNumber]%2 == 0){
        self.imageNumber = imageNumber+1;
        if (self.imageNumber > NUMBER_OF_IMAGES) {
            self.imageNumber = 1;
        }
        self.needsImageUpdated = YES;
    } else {
        self.needsImageUpdated = NO;
    }

    CGPoint newCenter = self.center;
    newCenter.y += self.speed;
    self.center = newCenter;

    if (newCenter.y - self.radius > controller.gameAreaSize.height){
        [controller removeActor: self];
    }
}
```

In Listing 5–12, we added two new properties to the asteroid class: `imageNumber` and `imageVariant`. The property `imageNumber` keeps track of the current image that should be displayed. Like the previous example, there are three types of asteroids: A, B, and C. The property `imageVariant` records which one of these sequences of images should be used.

In Listing 5–12, we can see that we added a new task called `imageName`. This task overrides the synthetic task as defined by `Actor03`. In this way, we alter the name of the image that is returned. We take the `imageVariant` and append the image number to it, creating a string that is of the form "Asteroid_B_0004" if the asteroid is of the B variety and is `imageNumber` is equal to 4.

Looking at the task step in Listing 5–12, we see that we added a new section at the beginning that updates which image should be displayed. Every other frame we increase the value if `imageNumber` by one, resetting back to one when it exceeds the constant `NUMBER_OF_IMAGES`. Each time we change `imageNumber` we want to set `needsImageUpdated` to YES, otherwise we set it to NO. Let's take a look at the task `updateActorView:` and see

Example03Controller uses this information to make sure the correct image is displayed. See Listing 5–22.

Listing 5–22. *Example03Controller.m (updateActorView:)*

```
-(void)updateActorView:(Actor03*)actor{
    UIImageView* imageView = [actorViews objectForKey:[actor actorId]];

    if (imageView == nil){
        UIImageView* imageView = [[UIImageView alloc] initWithImage:[UIImage
imageNamed:[actor imageName]]];
        [actorViews setObject:imageView forKey:[actor actorId]];
        [imageView setFrame:CGRectMake(0, 0, 0, 0)];
        [actorView addSubview:imageView];
    } else {
        if ([actor needsImageUpdated]){
            [imageView setImage:[UIImage imageNamed:[actor imageName]]];
        }

    }

    float xFactor = actorView.frame.size.width/self.gameAreaSize.width;
    float yFactor = actorView.frame.size.height/self.gameAreaSize.height;

    float x = (actor.center.x-actor.radius)*xFactor;
    float y = (actor.center.y-actor.radius)*yFactor;
    float width = actor.radius*xFactor*2;
    float height = actor.radius*yFactor*2;
    CGRect frame = CGRectMake(x, y, width, height);

    imageView.transform = CGAffineTransformIdentity;
    [imageView setFrame:frame];
    imageView.transform = CGAffineTransformRotate(imageView.transform, [actor
rotation]);

}
```

In Listing 5–22, we update the image of the imageView if the property needsImageUpdated is set to True by simply calling imageName on the actor again. That's all it takes to make the asteroids appear to roll through space.

The Rotating Effect

Now that we have way to indicate that the image of an actor should be updated, we can use that same feature to give some life to the spaceship. At the bottom of Listing 5–22 we can see that the transformation of the imageView is first modified to an identify transformation, then its frame is set, and finally the transform is modified again, applying the rotation.

The addition of the rotation logic in Listing 5–22 allows us to create a spaceship that rotates toward its moveToPoint before traveling to it. Figure 5–13 shows the sprites we are going to use while the ship is in different states.

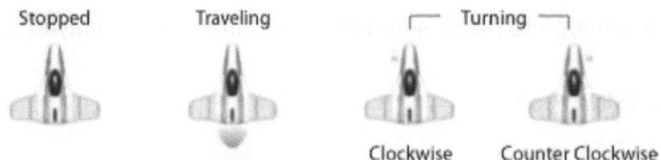

Clockwise   Counter Clockwise

Figure 5–13. *Images used for different spaceship states*

In Figure 5–13, we see four images. The image on the left is used when the spaceship is stopped. The second image from the left is used while the ship is in motion. The two images on the right are used when the ship is turning—one for each direction the ship will turn. Let's take a look at the header file for the class Viper03 and see what changes are required to make this work. See Listing 5–23.

Listing 5–23. *Viper03.h*

```
#define STATE_STOPPED 0
#define STATE_TURNING 1
#define STATE_TRAVELING 2

#import <Foundation/Foundation.h>
#import "Actor03.h"

@interface Viper03 : Actor03 {


}
@property CGPoint moveToPoint;
@property int state;
@property BOOL clockwise;

+(id)viper:(Example03Controller*)controller;
-(void)doCollision:(Actor03*)actor In:(Example03Controller*)controller;
@end
```

In Listing 5–23, we see that we define some constants representing our three states. We have also added a new state property to record the current state of the Viper03. We also added the property clockwise to keep track of which direction we are turning. So we know from the Asteroid03 class that we will be specifying which image we will be using in the task imageName, Listing 5–25 shows what the imageName task looks like for the class Viper03.

Listing 5–25. *Viper03.m (imageName)*

```
-(NSString*)imageName{
    if (self.state == STATE_STOPPED){
        return @"viper_stopped";
    } else if (self.state == STATE_TURNING){
        if (self.clockwise){
            return @"viper_clockwise";
        } else {
            return @"viper_counterclockwise";
        }
    } else {//STATE_TRAVELING
        return @"viper_traveling";
    }
}
```

In Listing 5–25, the task viper name simply returns the image name based on the state of the object. If the state is STATE_TURNING, we check the property clockwise to decide which image should be used. In order to implement the change in behavior we have to take a look at the new implementation of the task step:, as shown in Listing 5–24.

Listing 5–24. *Viper03.m (step:)*

```
-(void)step:(Example03Controller*)controller{
    CGPoint c = [self center];
    if (self.state == STATE_STOPPED){
        if (abs(moveToPoint.x - c.x) < self.speed && abs(moveToPoint.y - c.y) <
self.speed){
            c.x = moveToPoint.x;
            c.y = moveToPoint.y;
            [self setCenter:c];
        } else {
            self.state = STATE_TURNING;
            self.needsImageUpdated = YES;
        }
    } else if (self.state == STATE_TURNING){
        float dx = (moveToPoint.x - c.x);
        float dy = (moveToPoint.y - c.y);
        float theta = -atan(dx/dy);

        float targetRotation;

        if (dy > 0){
            targetRotation = theta + M_PI;
        } else {
            targetRotation = theta;
        }

        if ( fabsf(self.rotation - targetRotation) < .1){
            self.rotation = targetRotation;
            self.state = STATE_TRAVELING;
            self.needsImageUpdated = YES;
            return;
        }

        if (self.rotation - targetRotation < 0){
            self.rotation += .1;
            self.clockwise = YES;
            self.needsImageUpdated = YES;
        } else {
            self.rotation -= .1;
            self.clockwise = NO;
            self.needsImageUpdated = YES;
        }

    } else {//STATE_TRAVELING
        float dx = (moveToPoint.x - c.x);
        float dy = (moveToPoint.y - c.y);
        float theta = atan(dy/dx);

        float dxf = cos(theta) * self.speed;
        float dyf = sin(theta) * self.speed;
```

```
        if (dx < 0){
            dxf *= -1;
            dyf *= -1;
        }

        c.x += dxf;
        c.y += dyf;

        if (abs(moveToPoint.x - c.x) < self.speed && abs(moveToPoint.y - c.y) <
self.speed){
            c.x = moveToPoint.x;
            c.y = moveToPoint.y;
            self.state = STATE_STOPPED;
            self.needsImageUpdated = YES;
        }

        [self setCenter:c];
    }
}
```

In Listing 5–24, we see that we now have a set of if statements that controls the ship's behavior based on its state. If the ship is stopped, we check to se if it should stay stopped; if not, we switch to the turning state and indicate that we need the image updated by setting needsImageUpdated to Yes.

If the ship is in the turning state, calculate the angle we are turning toward based moveToPoint and store the result in the variable theta. Because atan only returns a value between - π /2 and π /2, we test to see if we should add π to get our targetRotation. Once we have our targetRotation value we check to see how close the targetRotation is to the current rotation. If they are close, we simply set rotate and to targetRotation and change state to STATE_TRAVELING. If we have not yet reached our targetRotation value, we rotate the ship a little way toward our targetRotation. If the ship is traveling, it moves toward the moveToPoint just as it did before.

Summary

In this chapter, we explored how to create a frame-by-frame animation where the item or actor in the game updates its location incrementally, many times a second. You learned how to synchronize an animation loop with the refresh rate of the device screen, and how to describe the state of a game in a display-agnostic way, freeing you from the coordinate system of the host UIView. We continued our exploration of this type of game by looking at some simple techniques to create animations and adding life to our actors.

Chapter 6

Create Your Characters: Game Engine, Image Actors, and Behaviors

In this chapter and the next, we will be creating classes for characters that will be used in the complete game presented in this book. But first, we need to refactor the code from the previous chapters to create reusable classes that make up a simple game engine. We will explore the classes and protocols that make up this game engine so you can understand how they fit together.

We will further abstract how actors are drawn on the screen by introducing the protocol Representation, which will allow us to create image-based actors as well as actors that are programmatically drawn (presented in the following chapter).

We will also introduce the concept of behavior through the protocol Behavior. This protocol provides us with a pattern for describing what an actor does in the game in a reusable way.

By building our first character, the power-up, you will start to learn how to use a game engine. We will create an example GameController subclass that lets a bunch of power-ups loose on the screen, so you can get a feel for how this character looks and behaves.

Understanding the Game Engine Classes

Our simple game engine starts with two core classes: GameController and Actor. GameController is very much like the controller classes from the preceding chapter, except that is has been refined for more-general use. The Actor class has similarly been improved to accommodate a greater number of Actor types. The Actor class also defines two protocols, Representation and Behavior, that are used to describe each actor in different ways. The protocol Representation describes the tasks required to create and update an actor's UIView. The protocol Behavior describes a task to be implemented by any class that wishes to create a shared, reusable behavior. These classes and protocols are covered in detail in the following sections.

The GameController Class

The GameController class coordinates the actors in the game and is ultimately responsible for rendering each actor on the screen. This includes setting up GameController as a UIViewController, defining a way to call our updateScene task repeated with a CADisplayLink, and finally reviewing how updateScene manages the actors in the scene. GameController also provides a mechanism for adding and removing actors. Let's take a look at the header for the class GameController, shown in Listing 6–1.

Listing 6–1. *GameController.h*

```
#import <UIKit/UIKit.h>
#import <QuartzCore/CADisplayLink.h>
#import "Actor.h"

@interface GameController : UIViewController {
    IBOutlet UIVicw* actorsView;

    CADisplayLink* displayLink;

    NSMutableSet* actors;
    NSMutableDictionary* actorClassToActorSet;

    NSMutableSet* actorsToBeAdded;
    NSMutableSet* actorsToBeRemoved;

    BOOL workComplete;
}
@property (nonatomic) long stepNumber;
@property (nonatomic) CGSize gameAreaSize;
@property (nonatomic) BOOL isSetup;
@property (nonatomic, retain) NSMutableArray* sortedActorClasses;

-(BOOL)doSetup;
-(void)displayLinkCalled;
-(void)updateScene;

-(void)removeActor:(Actor*)actor;
-(void)addActor:(Actor*)actor;
-(void)updateViewForActor:(Actor*)actor;

-(void)doAddActors;
-(void)doRemoveActors;
-(NSMutableSet*)actorsOfType:(Class)class;
@end
```

As you can see, the header for the class GameController extends the class UIViewController. The UIView named actorsView is an IBOutlet and is the UIView that contains the views for each actor in the game. We don't want to use the property view as the root view for the actors, because we may want to have other views associated with a GameController, such as a background image or other UIViews placed on top of the game. We will take advantage of this setup in a future chapter. For now, all you need to know is that the UIView actorsView is a subview of the property view and will contain all of the actor views.

In this listing, you also can see the now familiar CADisplayLink as well as various collections. The NSMutableSet actor stores all of the actors in the game. The NSMutableDictionary actorClassToActorSet is used to keep track of specific types of actors. The two NSMutableSets, actorsToBeAdded and actorsToBeRemoved, are used to keep track of actors that are created during a single step of the game and actors that should be removed after the current step.

The last field declared is the BOOL workComplete. This is used for debugging and will be explained later in this chapter, when we look at how the CADisplayLink is used. In addition to the fields declared in Listing 6–1, we also see four properties. The first property, stepNumber, keeps track of the number of steps that have passed since the game started. CGSize gameAreaSize is the size of the game area in game coordinates. The BOOL isSetup keeps track of whether the GameController has been set up yet. The last property, sortedActorClasses, is used to tell the GameController which types of actors it should keep track of.

Setting Up GameController

The use of all of the fields and properties in Listing 6–1 will be made clear as you inspect the implementation of the class GameController. Let's start with the task doSetup, shown in Listing 6–2.

Listing 6–2. *GameController.m (doSetup)*

```
-(BOOL)doSetup
{
    if (!isSetup){
        gameAreaSize = CGSizeMake(1024, 768);

        actors = [NSMutableSet new];

        actorsToBeAdded = [NSMutableSet new];
        actorsToBeRemoved = [NSMutableSet new];

        stepNumber = 0;

        workComplete = true;
        displayLink = [CADisplayLink displayLinkWithTarget:self
selector:@selector(displayLinkCalled)];
        [displayLink addToRunLoop:[NSRunLoop currentRunLoop]
forMode:NSDefaultRunLoopMode];
        [displayLink setFrameInterval:1];
        isSetup = YES;

        return YES;
    }
    return NO;
}
```

Here you can see the setup code for the class GameController. Because we want to set up a GameController only once, we put the setup code inside an if statement governed by the variable isSetup. The isSetup task will return YES if setup was performed, and return NO if setup was not performed.

As for the actual setup of the class, we have to initialize several variables. We start by setting the property gameSizeArea to a default size. Then we initialize the collection's actors, actorsToBeAdded, and actorsToBeRemoved. Finally, we set the stepNumber to zero.

Calling displayLinkCalled and updateScene

In Listing 6–2, the last step required is to set up the CADisplayLink. In this chapter, we set up the CADisplayLink slightly differently from previous examples. Instead of having the CADisplayLink call updateScene directly, we have it call displayLinkCalled, which in turn calls updateScene, as shown in Listing 6–3.

Listing 6–3. *GameController.m (displayLinkCalled)*

```
-(void)displayLinkCalled{
  if (workComplete){
      workComplete = false;
    @try {
        [self updateScene];
        workComplete = true;
    }
    @catch (NSException *exception) {
        NSLog(@"%@", [exception reason]);
        NSLog(@"%@", [exception userInfo]);//break point here
    }
  }
}
```

The task displayLinkCalled calls updateScene only if the variable workComplete is true. The variable workComplete can be true only if this is the first time displayLinkCalled is called or if updateScene has completed successfully. If updateScene throws an exception, it is logged, and workComplete is not reset to true. This, in effect, causes the game to stop if there are any errors. This is desirable because exceptions raised by tasks called by CADisplayLink are silently swallowed. To facilitate debugging the task displayLinkCalled, it is required to both log the exception and halt the game, letting the tester know that something has gone wrong. In production, you may wish to change this behavior; it might be better to keep the app moving even if there was an error. The exact behavior required will depend on the application.

Updating updateScene

The task updateScene is also updated from the preceding chapter, as shown in Listing 6–4.

Listing 6–4. *GameController.m (updateScene)*

```
-(void)updateScene{
    for (Actor* actor in actors){
        [actor step:self];
    }
    for (Actor* actor in actors){
        for (NSObject<Behavior>* behavoir in [actor behaviors]){
            [behavoir applyToActor:actor In:self];
        }
    }
    for (Actor* actor in actors){
        [self updateViewForActor:actor];
    }
    [self doAddActors];
    [self doRemoveActors];
    stepNumber++;
}
```

In this listing, we iterate over all of the actors in the game three times. In the first loop, we call step: on each actor, giving it a chance to apply any custom code. In the second loop, we apply each Behavior associated with the actor. Behavior is a protocol that describes some shared behavior among actors (this protocol is defined in Actor.h, shown later in Listing 6–9). In the third loop, we update the UIView representing each actor by calling updateViewForActor:.

Because actors and behaviors are free to add or remove other actors and we don't want to modify the NSMutableSet actors while we are iterating over that same collection, we have to add and remove actors in two steps. To implement this two-step process, we store newly added actors in the NSMutableSet actorsToBeAdded and then process each actor in that array in the task doAddActors. We follow an identical pattern for actors to be removed, storing them in the NSMutableSet actorsToBeRemoved and then processing them in the task doRemoveActors.

Calling doAddActors and doRemoveActors

The last step we take in updateScene is to call doAddActors and doRemoveActors, as shown in Listing 6–5.

Listing 6–5. *GameController.m (doAddActors and doRemoveActors)*

```
-(void)doAddActors{
    for (Actor* actor in actorsToBeAdded){
        [actors addObject:actor];

        UIView* view = [[actor representation] getViewForActor:actor In:self];
        [view setFrame:CGRectMake(0, 0, 0, 0)];
        [actorsView addSubview:view];
```

```
            NSMutableSet* sorted = [actorClassToActorSet valueForKey:[[actor class]
    description]];
            [sorted addObject:actor];
        }
    [actorsToBeAdded removeAllObjects];
}
-(void)doRemoveActors{
    for (Actor* actor in actorsToBeRemoved){

        UIView* view = [[actor representation] getViewForActor:actor In:self];
        [view removeFromSuperview];

        NSMutableSet* sorted = [actorClassToActorSet valueForKey:[[actor class]
    description]];
        [sorted removeObject:actor];

        [actors removeObject:actor];
    }
    [actorsToBeRemoved removeAllObjects];
}
```

The task doAddActors iterates through all actors in the NSMutableSet actorsToBeAdded and adds each one to the NSMutableSet actors. Additionally, a UIView for each actor is retrieved from the actor's representation property and added as a subview of actorsView. The property representation is an NSObject that conforms to the protocol Representation (as defined in Actor.h in Listing 6–9). After adding the view, we find the NSMutableSet sorted that corresponds to the actor's class in the NSMutableDictionary actorClassToActorSet. The actor is then added to the NSMutableSet sorted. In this way, we can later find all actors of a particular type with a quick lookup in the NSMutableDictionary actorClassToActorSet. The last step in doAddActors is to remove all of the actors from the set actorsToBeRemoved.

In the preceding listing, we see the task doRemoveActors that parallels the task doAddActors. In doRemoveActors, each actor in the NSMutableSet actorsToBeRemoved has its associated UIView removed from its super view and is removed from the corresponding NSMutableSet sorted. Each actor is also removed from the NSMutableSet actors. Finally, the NSMutableSet actorsToBeRemoved is cleared of all actors by calling removeAllObjects.

Adding and Removing Actors

To add or remove an actor from the game, we call addActor or removeActor, respectively, as shown in Listing 6–6.

Listing 6–6. *GameController.m (addActor and removeActor)*

```
-(void)addActor:(Actor*)actor{
    [actor setAdded:YES];
    [actorsToBeAdded addObject:actor];
}
-(void)removeActor:(Actor*)actor{
    [actor setRemoved:YES];
    [actorsToBeRemoved addObject:actor];
}
```

In this listing, you can see two very simple tasks: addActor and removeActor. In addActor, the passed-in actor has its added property set to YES before the actor is added to the NSMutableSet actorsToBeAdded. Similarly, in the task removeActor, the passed-in actor has its removed property set to YES and is added to the NSMutableSet actorsToBeRemoved.

Earlier, in Listing 6–5, when actors are added or removed from the set actors, we also added or removed the actors from sets stored in the NSMutableDictionary actorClassToActorSet. This is done so we can find all actors of a given type without iterating through the entire set actors. We don't, however, want to keep a separate set for each type of actor in the game. We just want to keep track of the types of actors we will need to access in the game logic.

Sorting Actors

The task setSortedActorClasses is used to specify which actors should be sorted, as shown in Listing 6–7.

Listing 6–7. *GameController.m (setSortedActorClasses:)*

```
-(void)setSortedActorClasses:(NSMutableArray *)aSortedActorClasses{
    [sortedActorClasses removeAllObjects];
    [sortedActorClasses release];
    sortedActorClasses = aSortedActorClasses;

    [actorClassToActorSet removeAllObjects];
    [actorClassToActorSet release];
    actorClassToActorSet = [NSMutableDictionary new];

    for (Class class in sortedActorClasses){
        [actorClassToActorSet setValue: [NSMutableSet new] forKey:[class description]];
    }

    for (Actor* actor in actors){
        NSMutableSet* sorted = [actorClassToActorSet objectForKey:[[actor class]
description]];
        [sorted addObject:actor];
    }

}
```

We start with a little memory management, removing all objects from the NSMutableArray sortedActorClasses and releasing it before reassigning the sortedActorClasses to the passed-in NSMutableArray aSortedActorClasses. Next, we clear and release the old NSMutableDictionary actorClassToActorSet. Once the bookkeeping is done, we add a new NSMutableSet for each class in sortedActorClass to actorClassToActorSet. Finally, we iterate through all actors and add each one to the corresponding NSMutableSet. This last step is required to allow setSortedActorClasses to be called multiple times in an application, but considering this is a relatively expensive operation, it is best to call setSortedActorClasses once at the beginning of the game.

Managing the UIView

The last improvement to the GameController class is the task updateViewForActor:, shown in Listing 6–8.

Listing 6–8. *GameController.m (updateViewForActor:)*

```
-(void)updateViewForActor:(Actor*)actor{
    NSObject<Representation>* rep = [actor representation];

    UIView* actorView = [rep getViewForActor:actor In:self];
    [rep updateView:actorView ForActor:actor In:self];

    float xFactor = actorsView.frame.size.width/self.gameAreaSize.width;
    float yFactor = actorsView.frame.size.height/self.gameAreaSize.height;

    float x = (actor.center.x-actor.radius)*xFactor;
    float y = (actor.center.y-actor.radius)*yFactor;
    float width = actor.radius*xFactor*2;
    float height = actor.radius*yFactor*2;
    CGRect frame = CGRectMake(x, y, width, height);

    actorView.transform = CGAffineTransformIdentity;
    [actorView setFrame:frame];
    actorView.transform = CGAffineTransformRotate(actorView.transform, [actor
rotation]);

    [actorView setAlpha:[actor alpha]];
}
```

The task updateViewForActor: is responsible for managing the UIView associated with each actor. In the first line of this task, we get the actor's representation property and store it in the variable rep. The variable rep is an NSObject that conforms to the protocol Representation. The protocol Representation is defined in Actor.h (shown later in Listing 6–9) and describes how a UIView for a given actor is created and updated. UIViews are created when getViewForActor:In: is called. When a UIView is updated to reflect a change to an actor, the task updateView:ForActor:In: is called. These two tasks are discussed later in this chapter.

After the UIView actorView is identified, we calculate the region of its parent view that it should take up. The details of this calculation are described in Chapter 5. We add two new features, however; we rotate actorView based on the actor's rotation property, and we also set the alpha value of actorView, allowing actors to change their opacity during a game.

Now that we have looked at the class GameController and how it manages the actors in a game, let's look at the class Actor so you can understand how it allows for many different types of actors in any game.

The Actor Class

The class Actor is the super class for all actors in your game. It primarily provides information about an actor's location, but it also describes protocols that indicate how an actor behaves and is represented on the screen. In this section, we will look at how Actor is implemented, followed by some concrete examples. Let's start by taking a look at the header file of the Actor class, shown in Listing 6–9.

Listing 6–9. *Actor.h*

```
#import <Foundation/Foundation.h>

@class GameController, Actor;

@protocol Representation
-(UIView*)getViewForActor:(Actor*)anActor In:(GameController*)aController;
-(void)updateView:(UIView*)aView ForActor:(Actor*)anActor
In:(GameController*)aController;
@end

@protocol Behavior
-(void)applyToActor:(Actor*)anActor In:(GameController*)gameController;
@end

long nextId;
@interface Actor : NSObject {

}
//State
@property (nonatomic, retain) NSNumber* actorId;
@property (nonatomic) BOOL added;
@property (nonatomic) BOOL removed;

//Geometry
@property (nonatomic) CGPoint center;
@property (nonatomic) float rotation;
@property (nonatomic) float speed;
@property (nonatomic) float radius;

//Behavoir
@property (nonatomic, retain) NSMutableArray<Behavior>* behaviors;

//Representation
@property (nonatomic) BOOL needsViewUpdated;
@property (nonatomic, retain) NSObject<Representation>* representation;
@property (nonatomic) int variant;
@property (nonatomic) int state;
@property (nonatomic) float alpha;

-(id)initAt:(CGPoint)aPoint WithRadius:(float)aRadius
AndRepresentation:(NSObject<Representation>*)aRepresentation;
-(void)step:(GameController*)controller;
-(BOOL)overlapsWith: (Actor*) actor;
-(void)addBehavior:(NSObject<Behavior>*)behavior;

+(CGPoint)randomPointAround:(CGPoint)aCenter At:(float)aRadius;

@end
```

Here the protocol Representation is defined. This protocol must be conformed to by any object that is to manage the UIView associated with an actor. The idea here is that there are at least two kinds of UIViews used to draw an actor: those based on images and those based on programmatic drawing. There can be other ways to implement a UIView that represents an actor, such as a UIView with subviews. We don't cover that example in this chapter, but we do look at image-based actors in this chapter and vector-based actors in the next.

Regardless of how an actor's UIView is implemented, the class GameController is going to have to know how to get a reference to the correct UIView and give the Representation the chance to update how it looks. Hence, we have the two required tasks getViewForActor:In: and updateView:ForActor:In:.

In the preceding listing, we see a second protocol named Behavior. We are going to create some simple classes that describe a particular behavior of an actor that can be shared among different types of actors. For example, we will create a behavior called LinearMotion that describes how an actor can move across the screen. Several types of actors will use this behavior. This is where the protocol Behavior comes in; it provides us with a single required task called applyToActor:From: that will be called by a GameController for every step of the game.

The listing also shows a number of properties. Many of these should be familiar from the previous chapter—including center, speed, and radius. Some new properties have also been added to make the Actor class as reusable as possible. The two properties added and removed are used to keep track of whether the actor has been added or removed from game.

The property behaviors is an NSMutableArray containing objects that conform to the protocol Behavior. By adding Behavior objects to the property behaviors, we can customize how each actor behaves. This can be as simple as describing the actor's motion or sophisticated artificial intelligence. We provide a special task called addBehavior:, which is shorthand for getting the NSMutableArray behaviors and adding a Behavior object. The addBehavior: task also gives us a chance to lazy load the NSMutableArray behaviors so we don't create it if we don't need it.

You can also see in Listing 6–9 that we added a variant and state property. It is very common for actors to require these two properties, so they are included in the class Actor for simplicity. The last property is alpha, which is used to describe the transparency of the actor on the screen.

The tasks defined in Listing 6–9 are basically the same as the task described in the preceding chapter for the class Actor. The only real difference here is that when an actor is initialized, a Representation must be passed in. We have also added a utility task, randomPointAround:At:, which is used to create a random point on a circle of aRadius size around point aPoint.

Implementing Actor

The implementation of the class Actor is shown in Listing 6–10.

Listing 6–10. *Actor.m*

```
#import "Actor.h"

@implementation Actor
@synthesize actorId;
@synthesize added;
@synthesize removed;
@synthesize center;
@synthesize rotation;
@synthesize speed;
@synthesize radius;
@synthesize needsViewUpdated;
@synthesize representation;
@synthesize variant;
@synthesize state;
@synthesize alpha;
@synthesize behaviors;


-(id)initAt:(CGPoint)aPoint WithRadius:(float)aRadius
AndRepresentation:(NSObject<Representation>*)aRepresentation{
    self = [super init];
    if (self != nil){
        [self setActorId:[NSNumber numberWithLong:nextId++]];
        [self setCenter:aPoint];
        [self setRotation:0];
        [self setRadius:aRadius];
        [self setRepresentation:aRepresentation];
        [self setAlpha:1.0];
    }
    return self;
}
-(void)step:(GameController*)controller{
    //implemented by subclasses.
}
-(BOOL)overlapsWith: (Actor*) actor {
        float xdist = abs(self.center.x - actor.center.x);
        float ydist = abs(self.center.y - actor.center.y);
    float distance = sqrtf(xdist*xdist+ydist*ydist);
    return distance < self.radius + actor.radius;
}
-(void)setVariant:(int)aVariant{
    if (aVariant != variant){
        variant = aVariant;
        needsViewUpdated = YES;
    }
}
-(void)setState:(int)aState{
    if (aState != state){
        state = aState;
        needsViewUpdated = YES;
```

```
    }
}
-(void)addBehavior:(NSObject<Behavior>*)behavior{
    if (behaviors == nil){
        behaviors = [NSMutableArray new];
    }
    [behaviors addObject:behavior];
}
-(void)dealloc{
    [actorId release];
    [behaviors removeAllObjects];
    [behaviors release];
    [representation release];

    [super dealloc];
}

+(CGPoint)randomPointAround:(CGPoint)aCenter At:(float)aRadius{
    float direction = arc4random()%1000/1000.0 * M_PI*2;
    return CGPointMake(aCenter.x + cosf(direction)*aRadius, aCenter.y +
sinf(direction)*aRadius);
}
@end
```

At the top of this file, we see that we simply use @synthesize for each of the class's properties to create the getter and setter tasks, with the exception of the property state and variant. We define our own implementation for the setter task for the properties state and variant, so we can set needsViewUpdated to YES when these values change. This is done to ensure that the actor's Representation properly updates the UIView representing this actor when the actor's state or variation changes.

The task initAt:WithRadius:AndRepresentation: simply initializes the actor and sets the passed-in properties. We also set the alpha property to 1.0, so that actors are fully opaque by default.

Because an actor now retains a reference to a number of other objects, such as the representation and the collection of behaviors, we now have to implement a dealloc task for this class. As shown in Listing 6–10, we release all objects that are retained as well as remove all objects from the NSMutableArray behaviors before calling super dealloc.

Working with the Power-Up Actor

In the previous section, we looked at the classes GameController and Actor and learned that the Actor class requires a Representation object to be displayed. The Representation protocol describes what is required of a class to manage the UIView associated with an actor. In this section, we will explore the class ImageRepresentation and see how it uses images to represent an actor. In the preceding chapter, we used images to draw different actors. In this section, we will show how we can combine those techniques into a simple, reusable class called ImageRepresentation.

To illustrate ImageRepresentation, in our first example we will create power-ups and animate them across the screen, as shown in Figure 6–1.

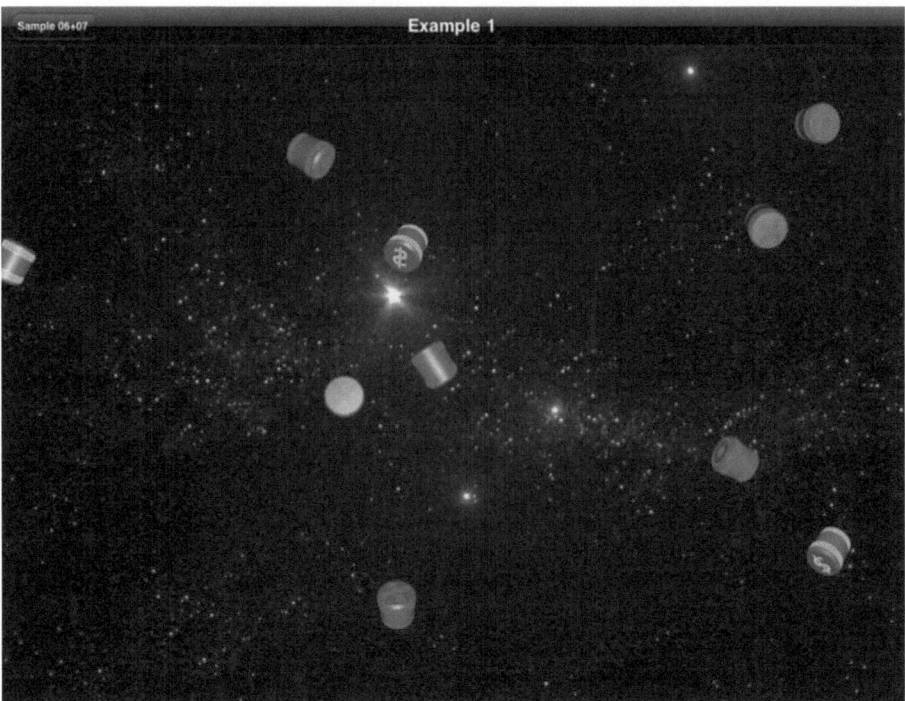

Figure 6–1. *Example 1—power-ups*

In the image, you can see several cylinders with three symbols: a dollar sign, a dot, and a cross. When this example is running, each cylinder travels in a straight line. When a cylinder reaches the edge of the screen, it wraps around to the opposite edge. As each cylinder travels, it also appears to spin or tumble about itself. After a few seconds, the bands and symbol on each cylinder will start flashing, announcing that it is about to disappear. When you start the example, there will be no power-ups on the screen, but every 5 seconds, a new one is added just offscreen and will drift into view. Let's take a closer look at the images used to create this animation. See Figure 6–2.

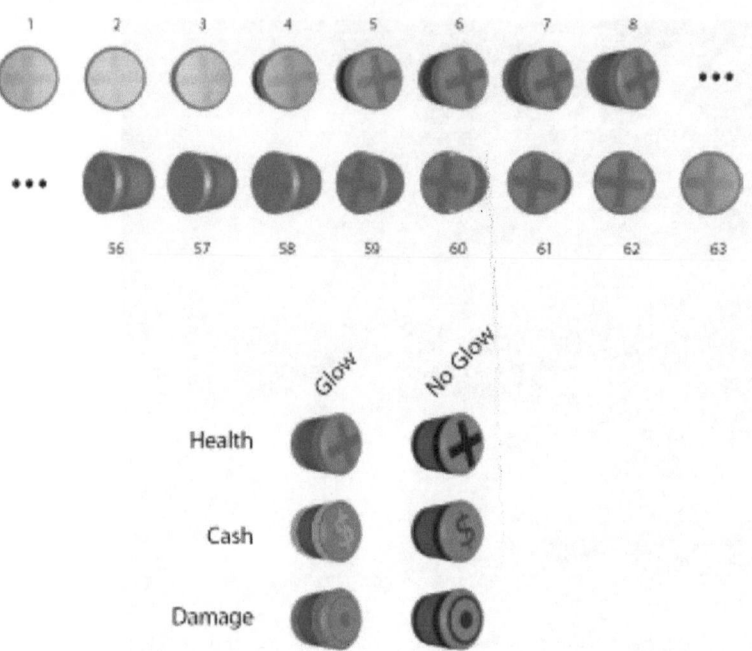

Figure 6–2. *Power-up detail*

The top of Figure 6–2 shows the beginning and the end of the 63 images that make up the rotating Health power-up with glow. In the example code, you will find six series of images: a glowing version and a nonglowing version for each of the three types of power-ups. By switching between the glowing version and the nonglowing version, we can create the blinking effect.

Implementing Our Power-Up Actor

To implement our power-up actor, we have to configure our GameController subclass, Example01Controller, to add power-ups to our scene. We also have to implement the power-up actor itself, giving it a way to draw itself and define its motion and behavior. The key classes are ImageRepresentation and Powerup, but we should start with the updateScene task from the class Example01Controller, as shown in Listing 6–11.

Listing 6–11. *Example01Controller.m (updateScene)*

```
-(void)updateScene{
    if (self.stepNumber % (60*5) == 0){
        [self addActor:[Powerup powerup: self]];
    }
    [super updateScene];
}
```

Here you can see the updateScene task as defined in the class Example01Controller. The class Example01Controller extends GameController so it inherits the game engine logic defined in that class. The updateScene task simply adds a new power-up every 5 seconds and then calls the super implementation of updateScene. The header file for the class Powerup is shown in Listing 6–12.

Listing 6–12. *Powerup.h*

```
#import <Foundation/Foundation.h>
#import "Actor.h"
#import "ImageRepresentation.h"
#import "ExpireAfterTime.h"

enum{
    STATE_GLOW = 0,
    STATE_NO_GLOW,
    PWR_STATE_COUNT
};

enum{
    VARIATION_HEALTH = 0,
    VARIATION_CASH,
    VARIATION_DAMAGE,
    PWR_VARIATION_COUNT
};

@interface Powerup : Actor <ImageRepresentationDelegate,ExpireAfterTimeDelegate>{

}

+(id)powerup:(GameController*)aController;

@end
```

This header file starts by defining two enumerations. The first enumeration defines the values for the two states of Powerup, STATE_GLOW or STATE_NO_GLOW. The second enumeration defines the values for the three variations of Powerup. Both of these enumerations end with a value that has the word COUNT in it. For those developers who are not coming from a C background, this is an old trick used to create a constant for the number of items in an enumeration. For example, in the first enumeration, STATE_GLOW has a value of 0, so STATE_NO_GLOW has a value of 1. This gives PWR_STATE_COUNT a value of 2, which is the number of states there are. Clever, no?

The class Powerup as defined in Listing 6–12 extends Actor. It also conforms to the protocols ImageRepresentationDelegate and ExpireAfterTimeDelegate. We will look at the tasks that come with these protocols when we look at the details of the ImageRepresentation and ExpireAfterTime. The only new task defined for the Powerup class is the constructor powerup:. Let's take a look at that task in Listing 6–13.

Listing 6–13. *Powerup.m (powerup: (partial))*

```
+(id)powerup:(GameController*)aController{

    CGSize gameSize = [aController gameAreaSize];
    CGPoint gameCenter = CGPointMake(gameSize.width/2.0, gameSize.height/2.0);
    float distanceFromCenter = sqrtf(gameCenter.x*gameCenter.x +
gameCenter.y*gameCenter.y);
    CGPoint center = [Actor randomPointAround:gameCenter At:distanceFromCenter];
```

```
    ImageRepresentation* rep = [ImageRepresentation imageRep];
    [rep setBackwards:arc4random()%2 == 0];
    [rep setStepsPerFrame:1 + arc4random()%3];


    Powerup* powerup = [[Powerup alloc] initAt:center WithRadius:32
AndRepresentation:rep];
    [rep setDelegate:powerup];
    float rotation = arc4random()%100/100.0 * M_PI*2;
    [powerup setRotation:rotation];
    [powerup setVariant:arc4random()%PWR_VARIATION_COUNT];

  //Section Omitted , see Listing 6-21

    return [powerup autorelease];
}
```

The first thing we have to do is figure out values that will be passed to Actor's
initAt:WithRadius:AndRepresentation task. After we have those values figured out, we
create a new power-up and then set a few starting values such as the rotation and
variation of the power-up. In an omitted section of Listing 6–13, we set up the behavior
of the power-up. This code can be found in Listing 6–21 and is discussed later in this
section.

Let's take a quick look at how we calculate the center of the power-up before diving into
the details of ImageRepresentation. The CGPoint center is created by picking a random
point on a circle just beyond the bounds of the screen, as shown in Figure 6–3.

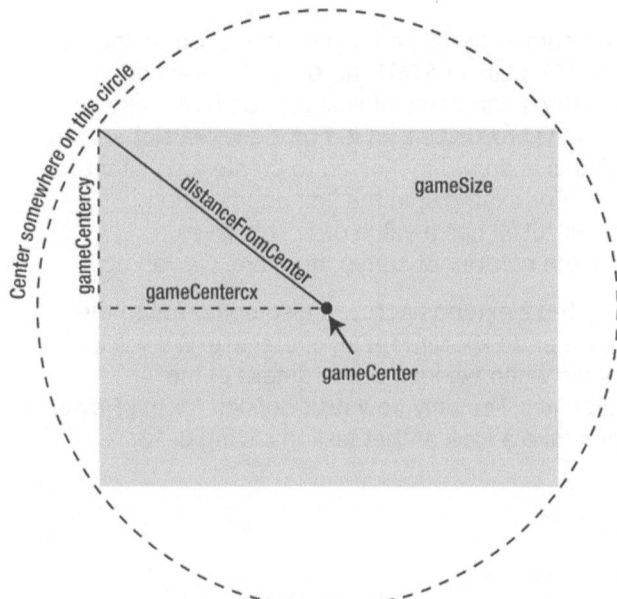

Figure 6–3. *Random point offscreen*

In this image, you can see the gray rectangle, which is the same size as the variable gameSize. The dot in the middle is the CGpoint gameCenter. The black line going from gameCenter to the upper-left corner of gameSize is of the length distanceFromCenter. The value of distanceFromCenter is calculated based on the x and y components of gameCenter, using the Pythagorean theorem. Using gameCenter and distanceFromCenter, we can get a random point on the outer circle by calling randomPointAround:At: from the class Actor (shown earlier in Listing 6–10).

In Listing 6–13, after calculating the center point and picking a radius of 32, we have to create a Representation object to handle the drawing of this power-up. Because we want to use a series of images to represent power-ups, we create an instance of ImageRepresentation called rep and set the properties backwards and stepsPerFrame. ImageRrepresentation is passed to Actor's initAt:WithRadius:AndRepresentation: task to set rep as the power-up's representation. Also note that we set the Powerup as the rep's delegate. This means that the instance of Powerup we are creating will be used to specify information about how to draw this actor.

Inspecting ImageRepresentation

Because we want to separate the game logic regarding an actor and how it is drawn, we are introducing the class ImageRepresentation. This class is responsible for creating the objects necessary for rendering an actor with a UIView. Primarily, ImageRepresentation will be created UIImageViews with the PNG files in order to draw our actors. Let's take a look at the header of ImageRepresentation, shown in Listing 6–14.

Listing 6–14. *ImageRepresentation.h*

```
#import <Foundation/Foundation.h>
#import "Actor.h"

@protocol ImageRepresentationDelegate

@required
-(NSString*)baseImageName;

@optional
-(int)getFrameCountForVariant:(int)aVariant AndState:(int)aState;
-(NSString*)getNameForVariant:(int)aVariant;
-(NSString*)getNameForState:(int)aState;

@end

@interface ImageRepresentation : NSObject<Representation> {
    UIView* view;
}
@property (nonatomic, assign) NSObject<ImageRepresentationDelegate>* delegate;
@property (nonatomic, retain) NSString* baseImageName;
@property (nonatomic) int currentFrame;
@property (nonatomic) BOOL backwards;
@property (nonatomic) int stepsPerFrame;

+(id)imageRep;
```

```
+(id)imageRepWithName:(NSString*)aBaseImageName;
+(id)imageRepWithDelegate:(NSObject<ImageRepresentationDelegate>*)aDelegate;
-(void)advanceFrame:(Actor*)actor ForStep:(int)step;
-(NSString*)getImageNameForActor:(Actor*)actor;
-(UIImage*)getImageForActor:(Actor*)actor;
```

@end

The goal of the ImageRepresentation class is to encapsulate code for handling the common ways images will be used to represent an actor. These include the following cases:

- A single image representing the actor

- A series of images representing the actor

- Actors with multiple states

- Actors with multiple variations

- Actors with a different number of images for every combination of states and variations

Each actor we create that uses an ImageRepresentation will fit one of the preceding cases. We indicate which one of these cases we intend by setting a delegate of type ImageRepresentationDelegate, as shown in Listing 6–14. This allows an instance of ImageRepresentationDelegate to indicate the name of the image that should be used at any given time.

The ImageRepresentationDelegate must specify a baseImageName, which is used in all of the use cases of ImageRepresentation. If the ImageRepresentationDelegate implements only baseImageName:, this indicates that a single image will be used for all instances of the actor. If a sequence of images is to be used, the delegate object must implement the task getFrameCountForVariant:AndState:. If a different sequence of images is to be used based on the state or variant of the actor, the delegate must implement getNameForState: and getNameForVariant:, respectively. Technically speaking, if just a single image is going to be used to represent all actors of a particular type, it is not required to specify a delegate to an ImageRepresentation. However, the ImageRepresentation still requires baseImageName to be specified.

Creating the Implementation of Powerup

The class Powerup must define how it is drawn as well as its basic behavior. Because the class Powerup will use a different sequence of images for each combination of state and variation, the delegate for ImageRepresentation must implement all four of these tasks. For simplicity, we specified that the class Powerup conforms to the protocol ImageRepresentationDelegate. Let's take a look at the implementation of these methods in Listing 6–15.

Listing 6–15. *Powerup.m (ImageRepresentationDelegate tasks)*

```
-(NSString*)baseImageName{
    return @"powerup";
}
-(int)getFrameCountForVariant:(int)aVariant AndState:(int)aState{
    return 63;
}
-(NSString*)getNameForVariant:(int)aVariant{
    if (aVariant == VARIATION_HEALTH){
        return @"health";
    } else if (aVariant == VARIATION_CASH){
        return @"cash";
    } else if (aVariant == VARIATION_DAMAGE){
        return @"damage";
    } else {
        return nil;
    }
}
-(NSString*)getNameForState:(int)aState{
    if (aState == STATE_GLOW){
        return @"glow";
    } else if (aState == STATE_NO_GLOW){
        return @"noglow";
    } else {
        return nil;
    }
}
```

In this listing, we see the four tasks required by an ImageRepresentationDelegate to create an actor represented by a different sequence of images for each combination of states and variations. The task baseImageName returns the NSString "powerup", indicating that all of the images for this actor will start with "powerup". The task getFrameCountForVariant:AndState: simply returns 63, because there are 63 images in for each combination of state and variation. Other types of actors may wish to have a different number of images for each combination of state and variation and are free to return different numbers in getFrameCountForVariant:AndState:.

The tasks getNameForVariant: and getNameForState: return string representations for each variation and state that apply to power-ups and allow ImageRepresentation to figure out exactly what image should be used at a given time.

Finding the Correct Image for an Actor

Let's take a look at ImageRepresentation's task getImageNameForActor: so you can understand how the four tasks from ImageRepresentationDelegate are used to identify an image name. See Listing 6–16.

Listing 6–16. *ImageRepresentation.m (getImageNameForActor:)*

```
-(NSString*)getImageNameForActor:(Actor*)actor{
    NSString* imageName = baseImageName;

    if (imageName == nil){
        imageName = [delegate baseImageName];
```

```
    }
    NSString* variant = nil;
    if ([delegate respondsToSelector:@selector(getNameForVariant:)]){
        variant = [delegate getNameForVariant:[actor variant]];
    }
    NSString* state = nil;
    if ([delegate respondsToSelector:@selector(getNameForState:)]){
        state = [delegate getNameForState:[actor state]];
    }
    int frameCount = 0;
    if ([delegate respondsToSelector:@selector(getFrameCountForVariant:AndState:)]){
        frameCount = [delegate getFrameCountForVariant:[actor variant] AndState:[actor
state]];
    }
    if (variant != nil){
        imageName = [[imageName stringByAppendingString:@"_"]
stringByAppendingString:variant];
    }
    if (state != nil){
        imageName = [[imageName stringByAppendingString:@"_"]
stringByAppendingString:state];
    }
    if (frameCount != 0){
        imageName = [[imageName stringByAppendingString:@"_"]
stringByAppendingString:[NSString stringWithFormat:@"%04d", currentFrame] ];
    }
    return imageName;
}
```

The goal here is to establish the correct name of the image that should be used to represent the actor. The final image name will be stored in the variable imageName. The first thing we do is establish the baseImageName. This can be either the baseImageName property of the ImageRepresentation or from the delegate.

The next step is to get the string representation of the actor's variation. We test to see whether the delegate responds to the task getNameForVariant. If it does, we assign variant to the result. We follow the same pattern to get a string representation for state.

The last thing we need the delegate for is to establish whether this actor is represented by a single image or by a sequence of images for this particular state and variation. If the delegate responds to getFrameCountForVariation:AndState:, we record the value in frameCount. If the delegate does not respond to getFrameCountForVariation:AndState:, or if getFrameCountForVariation:AndState: returned zero, this indicates that we should not include a frame number in the final image name.

After calling getFrameCountForVariation:AndState:, we have all of the information we require to establish the final name of the image. If the string representation of variant is not nil, we append "_" and variation to the NSString imageName. Similarly, if state is not nil, we append "_" and state to imageName. Finally, if frameCount is not zero, we append the currentFrame number to the image name. We don't have to append the file extension to the image name, because UIImage will automatically find the correct image based on the name. This task will find image files of the following patterns:

- baseImageName.png
- baseImageName_variant.png
- baseImageName_state.png
- baseImageName_variant_state.png
- baseImageName_OOFN.png
- baseImageName_state_OOFN.png
- baseImageName_variant_OOFN.png
- baseImageName_variant_state_OOFN.png

Now you know how ImageRepresentation and ImageRepresentationDelegate are used to specify which image should be used for a particular actor at a given time. Looking back at Listing 6–8, you can see that an actor's representation is responsible for creating and updating the UIView associated with the actor.

Creating the UIImageView for an Actor

The ultimate goal of ImageRepresentation is to create a UIView that represents the actor. This is done by instantiating a UIImageView that uses the correct image to represent the actor, given its type, state, and variation. We have reviewed how to find the correct image. All that is left is to create UIImageView. This work is done in the task getViewForActor:In:, shown in Listing 6–17.

Listing 6–17. *ImageRepresentation.m (getViewForActor:In:)*

```
-(UIView*)getViewForActor:(Actor*)actor In:(GameController*)aController{
    if (view == nil){
        UIImage* image = [self getImageForActor: actor];
        view = [[UIImageView alloc] initWithImage: image];
    }
    return view;
}
```

ImageRepresentation simply tests to see whether view has been created, and creates it if it has not. To create the UIImageView, we call getImageForActor: and then create UIImageView with the UIImage that was returned. The task getImageForActor: is shown in Listing 6–18.

Listing 6–18. *ImageRepresentation.m (getImageForActor:)*

```
-(UIImage*)getImageForActor:(Actor*)actor{
    NSString* imageName = [self getImageNameForActor:actor];

    UIImage* result = [UIImage imageNamed: imageName];
    if (result == nil){
        NSLog(@"Image Not Found: %@", imageName);
    }

    return result;
}
```

The task getImageForActor: simply calls getImageNameForActor and creates a UIImage based on that name. If no image exists with the given name, the result will be nil. Placing a beak point on that NSLog statement will save you a lot of trouble when debugging your app, as you will be notified immediately when an actor is being drawn with a nonexisting image.

Updating Our Views

Now that we have looked at how ImageRepresentation creates a UIView used to draw an actor, let's take a look at how ImageRepresentation updates that UIView to create the cycling animations and responds to a change in state or variation. Listing 6–19 shows the task updateView:ForActor:In: for ImageRepresentation.

Listing 6–19. *ImageRepresentation.m (updateView:ForActor:In:)*

```
-(void)updateView:(UIView*)aView ForActor:(Actor*)anActor
In:(GameController*)aController{

    if ([delegate respondsToSelector:@selector(getFrameCountForVariant:AndState:)]){
        [self advanceFrame: anActor ForStep:[aController stepNumber]];
    }

    if ([anActor needsViewUpdated]){

        UIImageView* imageView = (UIImageView*)aView;
        UIImage* image = [self getImageForActor: anActor];

        [imageView setImage:image];
        [anActor setNeedsViewUpdated:NO];
    }
}
```

The task updateView:ForActor:In: is responsible for making sure that the correct image is being used to represent the actor anActor. The first step is to see whether ImageRepresentation is managing a series of images by checking if the delegate responds to getFrameCountForVariation:AndState:. If a series of images is being used, we call advanceFrame:ForStep: to advance the current frame for the actor, as shown in Listing 6–20.

Listing 6–20. *ImageRepresentation.m (advanceFrame:ForTick:)*

```
-(void)advanceFrame:(Actor*)actor ForStep:(int)step{
    if (step % self.stepsPerFrame == 0){
        if (self.backwards){
            self.currentFrame -= 1;
        } else {
            self.currentFrame += 1;
        }

        int frameCount = [delegate getFrameCountForVariant:[actor variant]
AndState:[actor state]];
        if (self.currentFrame > frameCount){
            self.currentFrame = 1;
        }
        if (self.currentFrame < 1){
```

```
            self.currentFrame = frameCount;
        }

        [actor setNeedsViewUpdated:YES];
    }
}
```

We change the current frame only if the modulus of step by `stepsPerFrame` is zero. If backwards is true, we subtract 1 from the `currentFrame`; conversely, we add 1. If we change `currentFrame`'s value, we have to make sure we don't have a value less than 1 or greater than the number of frames. Note here that `currentFrame` is one based, not zero based. This was done so the value of `currentFrame` matches identically with our naming convention for image file names. This makes life easier when debugging. The last thing we do in `advanceFrame:ForStep:` is to mark this actor as needing its view updated.

Looking back at Listing 6–19, after calling `advanceFrame:ForStep:`, we check to see whether `needsViewUpdated` is true. If we had adjusted the `currentFrame` value in `advanceFrame:ForStep:`, then `needsViewUpdated` will be true. The value of `needsViewUpdated` could also be true for other reasons (perhaps the actor's state or variation has changed). Regardless of how `needsViewUpdates` was set to YES, we simply find the correct image for the actor by calling `getImageForActor:` and updating `imageView`. Finally, we set `needsViewUpdated` to NO, because we know at this moment that we have exactly the right image and we don't need to find a different image during the next step unless something sets `needsViewUpdated` back to YES.

We have reviewed the class `ImageRepresentation` and how it is used to coordinate the various states of the actor with the image intended to represent it. The goal in designing this class was to make it easy to facilitate creating new `Actor` classes that use it. In the other examples in the chapter, we will be reusing this class in various configurations and we will see how it handles these other cases.

The last thing to understand about this example is how the power-ups move across the screen and how they change state to create the blinking effect, as discussed in the following section.

Understanding Behaviors by Example

In this example, we have power-ups tumbling across the screen. Eventually they start blinking, just before they vanish. These two distinct behaviors could simply be implemented by providing the task `step:` in the class `Powerup`. In the previous chapter, this is how we implement the different behaviors of the different actors. In this chapter, we are introducing a general mechanism for defining behaviors that can be shared across different types of actors or changed based on a single actor's state. In Listing 6–13 earlier, we omitted the section of the class `Powerup`'s constructor that dealt with behaviors. Listing 6–21 shows this omitted code.

Listing 6–21. *Powerup.m (powerup: (partial))*

```
//Powerup* powerup = …
float direction = arc4random()%100/100.0 * M_PI*2;
LinearMotion* motion = [LinearMotion linearMotionInDirection:direction AndSpeed:1];
[motion setWrap:YES];
[powerup addBehavior: motion];

ExpireAfterTime* expire = [ExpireAfterTime expireAfter:60*30];
[expire setDelegate: powerup];
[powerup addBehavior: expire];

return [powerup autorelease];
```

We have just created a new Powerup object called powerup and set a few basic properties. The next step is to create Behavior objects and add them to the newly created Powerup before returning the objects. The first behavior we create is a LinearMotion object called motion, which we add to the powerup with the task addBehavior:. The second behavior is an ExpireAfterTime object called expire. Before adding expire to powerup, we register powerup as a delegate to expire. This gives us an easy way to make the power-up blink when it gets close to its expiration.

Behavior: Linear Motion

Both of these behaviors will be reused with other Actor classes, so they are worth looking at in some detail. Let's start with the class LinearMotion, whose header is shown in Listing 6–22.

Listing 6–22. *LinearMotion.h*

```
#import <Foundation/Foundation.h>
#import "Actor.h"

@interface LinearMotion : NSObject <Behavior>{
    float deltaX;
    float deltaY;
}

@property (nonatomic) float speed;
@property (nonatomic) float direction;
@property (nonatomic) BOOL wrap;

+(id)linearMotionInDirection:(float)aDirection AtSpeed:(float)aSpeed;
+(id)linearMotionRandomDirectionAndSpeed;

@end
```

The class LinearMotion extends NSObject and conforms to the protocol Behavior, which was imported from Actor.h. The class LinearMotion has three properties: speed, direction, and wrap. The properties speed and direction are hopefully obvious. The wrap property indicates whether the actor should reenter the game area on the opposite side after leaving the game area. The two tasks are constructors for creating LinearMotion objects. We could probably think of a few more handy constructors for this class, but these two are all we need for this chapter. Let's start exploring the implementation of the class LinearMotion by looking at these two constructors, shown in Listing 6–23.

Listing 6–23. *LinearMotion.m (constructors)*

```
+(id)linearMotionInDirection:(float)aDirection AtSpeed:(float)aSpeed{
    LinearMotion* motion = [LinearMotion new];
    [motion setDirection:aDirection];
    [motion setSpeed:aSpeed];

    return [motion autorelease];
}
+(id)linearMotionRandomDirectionAndSpeed{
    float direction = (arc4random()%100/100.0)*M_PI*2;
    float speed = (arc4random()%100/100.0)*3;

    return [LinearMotion linearMotionInDirection:direction AtSpeed:speed];
}
```

The constructor linearMotionInDirection:AtSpeed: simply creates a new LinearMotion object and sets the two properties. The constructor linearMotionRandomDirectionAndSpeed creates random values for direction and speed and then simply calls the constructor linearMotionInDirection:AtSpeed:.

Listing 6–24 shows the rest of the implementation of the class LinearMotion.

Listing 6–24. *LinearMotion.m*

```
-(void)setSpeed:(float)aSpeed{
    speed = aSpeed;
    deltaX = cosf(direction)*speed;
    deltaY = sinf(direction)*speed;
}
-(void)setDirection:(float)aDirection{
    direction = aDirection;
    deltaX = cosf(direction)*speed;
    deltaY = sinf(direction)*speed;
}

-(void)applyToActor:(Actor*)anActor In:(GameController*)gameController{
    CGPoint center = anActor.center;
    center.x += deltaX;
    center.y += deltaY;

    if (wrap){
        CGSize gameSize = [gameController gameAreaSize];
        float radius = [anActor radius];

        if (center.x < -radius && deltaX < 0){
            center.x = gameSize.width + radius;
        } else if (center.x > gameSize.width + radius && deltaX > 0){
            center.x = -radius;
        }

        if (center.y < -radius && deltaY < 0){
            center.y = gameSize.height + radius;
        } else if (center.y > gameSize.height + radius && deltaY > 0){
            center.y = -radius;
```

```
            }
        }
        [anActor setCenter:center];
}
```

We calculate the values deltaX and deltaY each time the speed or direction values are set. The task applyToActor:In: is defined by the protocol Behavior and is called by a GameController to give a Behavior a chance to modify an actor. In the implementation of applyToActor:In: for the class LinearMotion, we start by getting the current center of the actor and increase the x and y components by deltaX and deltaY.

If wrap is true, we have to see whether the actor is outside the game area. This is a little more complex than simply checking whether center is inside gameArea or not, because we have to take into account the radius of the actor as well. We want an actor to appear as though it is traveling offscreen instead of simply vanishing when it touches the edge of the screen. Also, we want to move an actor only if it is moving away from the game area. Keeping these two things in mind, we test whether the x value of center is less than radius and has a negative deltaX. This means the actor is to the left of the game area and moving left, away from the game area. So we had better move it to the right of the game area, so its leftward motion brings the actor back onto the right side of the screen. We perform the opposite operation if the x value of center is greater than the width of the game area and is moving to the right, moving the actor to the left side of the screen. These two tests are repeated for the center's y value.

The implementation of LinearMotion is pretty simple, only slightly complicated by the notion of wrapping. Next, we will explore the class ExpireAfterTime, which is used to remove actors from the scene after a given time.

Behavior: ExpireAfterTime

The class ExpireAfterTime is used by Powerup to remove it from the scene after a number of steps have passed by. Before the power-up is removed, we want to make the power-up blink by changing its state from glowing to not glowing. To support this second feature, we will want to have some sort of communication between the ExpireAfterTime object and the actor it is acting on. Listing 6–25 shows the header of the class ExpireAfterTime.

Listing 6–25. *ExpireAfterTime.h*

```
#import <Foundation/Foundation.h>
#import "Actor.h"

@class ExpireAfterTime;
@protocol ExpireAfterTimeDelegate
-(void)stepsUpdated:(ExpireAfterTime*)expire In:(GameController*)controller;
@end


@interface ExpireAfterTime : NSObject <Behavior> {

}
@property (nonatomic) long stepsRemaining;
```

```
@property (nonatomic, assign) Actor<ExpireAfterTimeDelegate>* delegate;

+(id)expireAfter:(long)aNumberOfSteps;
@end
```

Here the class ExpireAfterTime conforms to the protocol Behavior, just like the class
LinearMotion does. ExpireAfterTime also defines a new protocol called
ExpireAfterTimeDelegate that defines the task stepsUpdated:In:, which will be called
by the ExpireAfterTime on its delegate every time applyToActor:In: is called, as shown
in Listing 6–26.

Listing 6–26. *ExpireAfterTime.m*

```
#import "ExpireAfterTime.h"
#import "GameController.h"

@implementation ExpireAfterTime
@synthesize stepsRemaining;
@synthesize delegate;

+(id)expireAfter:(long)aNumberOfSteps {
    ExpireAfterTime* expires = [ExpireAfterTime new];
    [expires setStepsRemaining: aNumberOfSteps];
    return [expires autorelease];
}

-(void)applyToActor:(Actor*)anActor In:(GameController*)gameController{
    stepsRemaining--;

    [delegate stepsUpdated:self In:gameController];

    if (stepsRemaining <= 0){
        [gameController removeActor:anActor];
    }
}
@end
```

In this implementation of the class ExpireAfterTime, the constructor expireAfter:
simply creates a new ExpireAfterTime object and sets the stepsRemaining property. The
task applyToActor:In: is called by a GameController for each step of the game. In this
task, we decrement stepsRemaining and update the delegate that the number of steps
was updated. If stepsRemaining is zero or less, we remove the actor from the
GameController.

Before we add ExpireAfterTime to the Powerup object, we set the Powerup object as the
delegate to ExpireAfterTime. This causes ExpireAfterTime to call stepsUpdated:In: on
the Powerup object. Listing 6-27 shows the implementation of stepsUpdated:In: for the
class Powerup.

Listing 6–27. *Powerup.m (stepsUpdated:In:)*

```
-(void)stepsUpdated:(ExpireAfterTime*)expire In:(GameController*)controller{
    long stepsRemaining = [expire stepsRemaining];
    if (stepsRemaining < 60*5){
        if (stepsRemaining % 25 == 0){
            if (self.state == STATE_GLOW){
```

```
                self.state = STATE_NO_GLOW;
            } else {
                self.state = STATE_GLOW;
            }
        }
    }
}
```

In this listing, we check how many steps are remaining for the ExpireAfterTime object. If that value is fewer than 300 steps (5 seconds at 60 frames per second), we want to apply our blinking logic. To make the power-up blink, we swap the state of Powerup from STATE_GLOW to STATE_NO_GLOW, or vice versa.

Changing the state is all that is required to change which image represents the Powerup actor, because (looking back at Listing 6–10) we know that setting the state of an actor sets its needViewUpdated property to YES. Further, if an actor's needViewUpdated property is YES and that actor is using an ImageRepresentation, that ImageRepresentation will fetch the correct image for the actor in the task updateView:In: (as shown earlier in Listing 6–19).

Summary

In this chapter, we looked at our base game engine classes GameController and Actor and learned how to use these classes to create a simple scene exercising our new actor, the power-up. We looked at the protocol Behavior, which provides us with a framework for creating reusable bits of game logic to be applied to our actors. We also consolidated the logic required to render an image-based actor into the class ImageRepresentation. This also set the stage for vector-based actors, which will be discussed in the following chapter.

Chapter 7

Build Your Game: Vector Actors and Particles

Core Graphics is a powerful 2D drawing library that is responsible for rendering large portions of iOS and OS X. In this chapter, we see how to use this library to draw actors in our game. The goal is to have actors that are drawn dynamically based on game state. To illustrate this, we'll create two example actors drawn with Core Graphics: a health bar that shows the amount of health remaining for an actor, and a bullet that is drawn in a particular color depending on how powerful it is. These two examples will illustrate how to use Core Graphics within the context of the simple game engine started.

We will accomplish this by creating a new class called VectorRepresentation, which is analogous to the class ImageRepresentation from the previous chapter. The class VectorRepresentation will be used to create a UIView to represent our actor, and be drawn with custom code using Core Graphics.

We'll also look at another popular technique used in games to create a compelling visual: the particle system. Stated simply, a particle system is anything in a game that generates lots of little graphics that, when composed on the screen, create an overall effect more interesting than the sum of its parts. Particle systems are used in games to create fire effects, water effects, and spell effects, to name a few.

In this chapter, we use particle systems to create comets that are composed of a large number of simple particle actors to give the comets a glowing, fluid feel. We are also going to use this technique to illustrate a little realism when an asteroid breaks apart.

The sample code for this chapter can be found in the Xcode project Sample 06+07.

Saucers, Bullets, Shields, and Health Bars

In this example, we will look at four new actors: saucers, bullets, shields, and health bars. The code for this section is found under the group Example 2, in the Xcode project Sample 06+07.

The health bar and bullet actors will be rendered programmatically with Core Graphics instead of using a pre-rendered image. The other actors—saucers and shields—exist in this example to provide a little context (got to have something to shoot with the bullets) and flesh out our example from the last chapter in terms of behaviors. Figure 7–1 shows the saucers, bullets, shields, and health bars in action.

Figure 7–1. *Saucers, bullets, shields, and health bars*

In Figure 7–1, we see a flying saucer in the middle of the screen. Coming from the right and traveling left are a number of circular bullets. The bullets come in three different sizes, which reflect their potential for damage. If a bullet collides with the saucer, a shield effect is added to the flying saucer and the health bar goes down, indicating the damage done. In this example, the bullets will keep coming until the saucer's health is zero, at which point it is removed and a new flying saucer is added. Figure 7–2 shows the three difference saucers used in this example, as well as the graphic used for the shield.

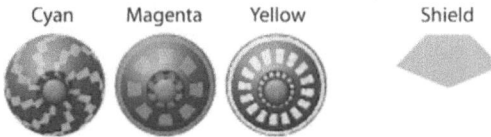

Cyan   Magenta   Yellow   Shield

Shield Over Saucer Scale x2

Figure 7–2. *Three saucers and a shield*

The shield image is partially transparent and is designed be drawn over a saucer, as can be seen in the scaled up saucer with the shield drawn over it.

To understand how all the parts and pieces of this example fit together, we will start by looking at the implementation of the class Example02Controller, starting with the doSetup task, shown in Listing 7–1.

Listing 7–1. *Example02Controller.m (doSetup)*

```
-(BOOL)doSetup
{
    if ([super doSetup]){
        NSMutableArray* classes = [NSMutableArray new];
        [classes addObject:[Saucer class]];
        [classes addObject:[Bullet class]];

        [self setSortedActorClasses:classes];
        return YES;
    }
    return NO;
}
```

In Listing 7–1, we see the doSetup task of the class Example02Controller. This method simply indicates that we want to keep actors of type Saucer and Bullet sorted for easy access later on. We do this by creating an NSMutableArray called classes and adding the class objects for two classes in which we are interested. The NSMutableArray classes are then passed to setSortedActorClasses:. This way, when we call actorsOfType: in the future, we will not have to sort through all of the actors to find them. Next, we want to take a look at the code that describes the action in this example that is found in the task updateScene of class Example02Controller. See Listing 7–2.

Listing 7–2. *Example02Controller.m (updateScene)*

```
-(void)updateScene{

    NSMutableSet* suacers = [self actorsOfType:[Saucer class]];

    if ([suacers count] == 0){
        Saucer* saucer = [Saucer saucer:self];
        [self addActor: saucer];
    }

    if (self.stepNumber % (30) == 0){
        float direction = M_PI - M_PI/30.0;
        float rand = (arc4random()%100)/100.0 * M_PI/15.0;
        direction += rand;
        Bullet* bullet = [Bullet bulletAt:CGPointMake(1000, 768/2)
WithDirection:direction];
        [self addActor: bullet];

        if (arc4random()%4 == 0){
            if (arc4random()%2 == 0){
                [bullet setDamage:20.0];
            } else {
                [bullet setDamage:30.0];
            }
        }
    }

    NSMutableSet* bullets = [self actorsOfType:[Bullet class]];

    for (Bullet* bullet in bullets){
        for (Saucer* saucer in suacers){
            if ([bullet overlapsWith:saucer]){
                [saucer doHit:bullet with:self];
            }
        }
    }
    [super updateScene];
}
```

In Listing 7–2, we first get all instances of Saucer in the scene and check to see if any exist. If none do, we create a new Saucer and add it to the scene with the addActor: task. This insures that there is always exactly one Saucer in the middle of the screen. Then, every 30 steps, we create a new Bullet using the task bullatAt:WithDirection:. The location of the bullet is specified in game coordinates, and the right-most point at the vertical center. The direction specified will cause the bullet to travel toward the left, possibly hitting the saucer. Once the bullet is added to the scene, the damage is randomly assigned.

The last thing the updateScene task does is check to see if any bullet has collided with the saucer. This is done by calling actorsOfType: and getting all Bullet objects in the scene stored in the NSMutableSet* bullets. We simply iterate over each bullet and saucer and check if there is a hit by calling overlapsWith:. If there is a hit, we call the doHit:with: task on saucer.

Let's take a closer look at the implementation of the actor classes used in this example.

The Actor Classes

In this section, we learn about the Saucer actor and the HealthBar Actor classes. The HealthBar is different from other Actors we have looked up until this point. It is different because its location in the scene is dependent on the location of another Actor, in this case, a Saucer. Once we have reviewed the classes Saucer and HealthBar, we will look at the Behavior class FollowActor and see how we implement this feature.

Instantiating the Saucer Class

We will start by looking at the constructor for the class Saucer. We want to start with the Saucer's constructor, because this is where we create the HealthBar. In this way, we don't have to worry about adding a HealthBar when we create a new Saucer, because one will be created automatically every time we create a Saucer. The constructor for Saucer is shown in Listing 7–3.

Listing 7–3. *Saucer.m (saucer:)*

```
+(id)saucer:(GameController*)controller{
    CGSize gameAreaSize = [controller gameAreaSize];
    CGPoint gameCenter = CGPointMake(gameAreaSize.width/2.0, gameAreaSize.height/2.0);

    ImageRepresentation* rep = [ImageRepresentation imageRep];
    [rep setBackwards:arc4random()%2 == 0];
    [rep setStepsPerFrame:3];

    Saucer* saucer = [[Saucer alloc] initAt:gameCenter WithRadius:32
AndRepresentation:rep];
    [rep setDelegate:saucer];

    [saucer setVariant:arc4random()%VARIATION_COUNT];
    [saucer setMaxHealth:100];
    [saucer setCurrentHealth:100];

    HealthBar* healthBar = [HealthBar healthBar:saucer];
    [healthBar setPercent:1];
    [saucer setHealthBar:healthBar];
    [controller addActor:healthBar];

    return saucer;
}
```

In Listing 7–3, we start by finding the center of the game area and storing that value in the CGPoint gameCenter. Next, we create an ImageRepresentation called rep. We indicate that rep should spin the saucer backward half the time at a rate of 3 steps per frame of animation. The CGRect gameCenter and the ImageRepresentation rep are passed to the super initializer initAt:WithRadius:AndRepresenation: to create a Saucer object with a radius of 32. The object saucer is then set as the delegate to the ImageRepresenation so we can specify the representational details with the class Saucer. This works exactly like the actors in Chapter 6. Have a look at the source code to see exactly how these tasks are implemented.

Once the saucer object is created, we want to initialize it with some details. We set which of the three variants we want to use by calling setVariant: on saucer. We also set the max health and current health properties to 100. Let's move on to the HealthBar class.

Instantiating the HealthBar Class

In order to render the health bar under the saucer, we create a HealthBar object and pass in the object we want it to follow, the actor saucer. We also set the healthBar property of saucer to the newly created healthBar so the saucer object can update the percentage displayed by the health bar when it takes damage. Lastly, healthBar is added to the scene. Let's take a look at the constructor for HealthBar and understand how it is set up and behaves the way it does. See Listing 7–4.

Listing 7–4. *HealthBar.m (healthBar:)*

```
+(id)healthBar:(Actor*)anActor{

    VectorRepresentation* rep = [VectorRepresentation vectorRepresentation];

    HealthBar* healthBar = [[HealthBar alloc] initAt:anActor.center
WithRadius:anActor.radius AndRepresentation:rep];
    [rep setDelegate:healthBar];

    [healthBar setColor:[UIColor blueColor]];
    [healthBar setBackgroundColor:[UIColor colorWithRed:0 green:0 blue:0 alpha:.5]];

    FollowActor* follow = [FollowActor followActor:anActor];
    [follow setYOffset:[anActor radius]];
    [healthBar addBehavior:follow];

    return healthBar;
}
```

In Listing 7–4, we see the constructor task healthBar: and that it takes an Actor as an argument. The Actor anActor is the actor that this HealthBar will stay under if it moves. In this example, the Saucer associated with it does not move, but we are going to use this class in future examples, where its actor does move. Before we look at the behavior FollowActor, notice that HealthBar is initialized by passing an instance of VectorRepresentation instead of an ImageRepresentation object. This indicates that we want to draw this object programmatically. We will look at how this works in the next section. For now, let's continue to explore this example. The next class we want to inspect is FollowActor.

The Behavior FollowActor Class

The class responsible for keeping the HealthBar near the Saucer is the class FollowActor. This class is a Behavior and provides the same sort of abstraction we found with the Behavior classes. Let's see how all of this fits together by looking at the class FollowActor. See Listing 7–5.

Listing 7–5. *FollowActor.m*

```
#import "FollowActor.h"
#import "GameController.h"

@implementation FollowActor
@synthesize actorToFollow;
@synthesize xOffset;
@synthesize yOffset;

+(id)followActor:(Actor*)anActorToFollow{
    FollowActor* follow = [FollowActor new];
    [follow setActorToFollow:anActorToFollow];
    return follow;
}
-(void)applyToActor:(Actor*)anActor In:(GameController*)gameController{
    if (![actorToFollow removed]){
        CGPoint c = [actorToFollow center];
        c.x += xOffset;
        c.y += yOffset;

        [anActor setCenter: c];
    } else {
        [gameController removeActor:anActor];
    }
}

@end
```

In Listing 7–5, we see the entire implementation of the class FollowActor. It is
constructed with the task followActor:, which takes the actor to follow as an argument.
The task applyToActor:In: is called for every step of the game and is responsible for
repositioning the actor this Behavior is associated with. To do this, we simply set the
center of the Actor anActor to center of the actor we are following, adjusting by xOffest
and yOffset. Looking at Listing 7–4, we see that the HealthBar is configured to follow
the saucer at a xOffset equal to the radius of saucer. This keeps the HealthBar
centered below saucer. Now let's explore the Bullet class.

The Bullet Class

We have seen how the Saucer and the HealthBar classes work together. We now want
to look at the Bullet class and understand why it behaves the way that it does. Listing
7–6 shows Bullet's constructor.

Listing 7–6. *Bullet.m (bulletAt:WithDirection:)*

```
+(id)bulletAt:(CGPoint)aCenter WithDirection:(float)aDirection{

    VectorRepresentation* rep = [VectorRepresentation vectorRepresentation];

    Bullet* bullet = [[Bullet alloc] initAt:aCenter WithRadius:4 AndRepresentation:rep];
    [rep setDelegate:bullet];

    [bullet setDamage: 10];
```

```
        LinearMotion* motion = [LinearMotion linearMotionInDirection:aDirection AtSpeed:3];
        [motion setWrap:YES];

        [bullet addBehavior: motion];

        ExpireAfterTime* expires = [ExpireAfterTime expireAfter:240];
        [bullet addBehavior:expires];

        return bullet;
}
```

In Listing 7–6, we see the class is composed of familiar building blocks. It is initialized with the now familiar task `initAt:WithRadius:AndRepresentation:`, and we are adding the `Behavior` objects of type `LinearMotion` and `ExpiresAfterTime`. These two `Behavior` objects are used by the class `Powerup` and are fully described in Chapter 6. The only new feature in the `Bullet` class is the class `VectorRepresentation`, which was also used by the class `HealthBar` to indicate that this type of actor will be drawn using Core Graphics. Let's take a closer look at the class `VectorRepresentation` and see how it interacts with our game engine and actors to provide a simple method for rendering our actors with Core Graphics.

Drawing Actors with Core Graphics via VectorRepresentation

The Core Graphics library is interesting because it is not a vector graphics library (like SVG), but it is excellent at rasterizing vector data structures. This makes a lot of sense if you consider that Core Graphics was originally based on PDF. Core Graphics is a great library for rendering high-quality graphics in a resolution-independent way. What this means for us is that our code will not be specifying which pixel is which color, but instead will be describing shapes and how they should be filled. Core Graphics is also known as Quartz. A complete description of the available drawing functions can he found here: `http://goo.gl/6i5gr`

In iOS a `UIView` can specify how it is drawn by overriding the task `drawRect:`, getting a reference to its graphics context and then calling a number drawing methods. Because our game engine uses `UIView`'s to represent each actor, we know that ultimately we will be creating a subclass of `UIView` and calling the drawing code for our vector-based actors. It is the job the class `VectorRepresentation` to manage the relationship between the actor and the view that draws it,

The VectorRepresentation Class

The `VectorRepresentation` class is responsible for creating a `UIView` suitable to draw our actor on. Let's start exploring this class with the header for `VectorRepresentation`. See Listing 7–7.

Listing 7–7. *VectorRepresentation.h*

```
#import <Foundation/Foundation.h>
#import "Actor.h"

@class VectorActorView;

@protocol VectorRepresentationDelegate
-(void)drawActor:(Actor*)anActor WithContext:(CGContextRef)context InRect:(CGRect)rect;
@end

@interface VectorRepresentation : NSObject<Representation> {
    VectorActorView* view;
}
@property (nonatomic, assign) NSObject<VectorRepresentationDelegate>* delegate;
+(id)vectorRepresentation;
@end
```

In Listing 7–7, we see that the class VectorRepresentation defines a protocol called VectorRepresentationDelegate to which each vector-based actor must conform. This protocol requires the task drawActor:WithContext:InRect: to be implemented by all classes that conform to it, and is where the actual drawing code should be placed.

The class VectorRepresentation extends NSObject and conforms to the protocol Representation, just like the class ImageRepresentation does. The idea here is that most of code should not care which type of representation we use—vector or image. We simply create the correct Representation object and associate it with each actor.

We also see that the class VectorRepresentation has a single property view of type VectorActorView. A VectorActorView is a subclass of UIView and will ultimately call the task drawActor:WithContext:InRect: as defined by the protocol VectorRepresentationDelegate. Before looking at the class VectorActorView, let's look at the implementation of VectorRepresentation as shown in Listing 7–8.

Listing 7–8. *VectorRepresentation.m*

```
#import "VectorRepresentation.h"
#import "VectorActorView.h"

@implementation VectorRepresentation
@synthesize delegate;

+(id)vectorRepresentation{
    return [VectorRepresentation new];
}
-(UIView*)getViewForActor:(Actor*)anActor In:(GameController*)aController{
    if (view == nil){
        view = [VectorActorView new];
        [view setBackgroundColor:[UIColor clearColor]];
        [view setActor: anActor];
        [view setDelegate:delegate];
        [anActor setNeedsViewUpdated:YES];
    }
    return view;
}
-(void)updateView:(UIView*)aView ForActor:(Actor*)anActor
In:(GameController*)aController{
```

```
        if ([anActor needsViewUpdated]){
            [aView setNeedsDisplay];
            [anActor setNeedsViewUpdated:NO];
        }
    }
@end
```

In Listing 7–8, we see the implementation of the class VectorRepresentation and that it has a trivial constructor. We also see that it implements the tasks getViewForActor:In: and updateView:In: as defined by the protocol Representation. In the task getViewForActor:In: we create a new VectorActorView, and specify the actor and delegate that should be used to draw. Then we call setNeedsViewUpdated, making sure the actor's drawing code is called at least once.

In the task, updateView:In: we check to see if the actor needs updating, if so we call setNeedsDisplay on the UIView aView. The task setNeedsDisplay is defined by the class UIView and indicates to the underlying graphics system that the UIView should be redrawn. In effect, this causes the drawRect: task of our VectorActorView to be called, which in turn will call the drawActor:WithContext:InRect: of our actor. Next we want to look at the details of the class VectorActorView.

A UIView for Vector-Based Actors: VectorActorView

The class VectorActorView is a subclass of UIView and is used to represent the actor in the scene in much the same way as we use UIImageView instances to represent image-based Actors. Let's take a look at the class VectorActorView and see how it all fits together, starting with the header. See Listing 7–9.

Listing 7–9. *VectorActorView.h*

```
#import <UIKit/UIKit.h>
#import "Actor.h"
#import "VectorRepresentation.h"

@interface VectorActorView : UIView {

}
@property (nonatomic, retain) Actor* actor;
@property (nonatomic, retain) NSObject<VectorRepresentationDelegate>* delegate;

@end
```

In Listing 7–9, we see that VectorActorView extends UIView and has two properties. The first is an Actor that is the actor to be drawn. The second property is the VectorRepresentationDelegate that is responsible for actually drawing the actor. In this example, these two objects—the actor and the delegate—will be the same object, either a Bullet or a HealthBar. The implementation of the class VectorActorView is very simple, as can be seen in Listing 7–10.

Listing 7–10. *VectorActorView.m*

```
- (void)drawRect:(CGRect)rect
{
    CGContextRef context = UIGraphicsGetCurrentContext();
    [delegate drawActor:actor WithContext:context InRect:rect];
}
```

In Listing 7–10, we see that the implementation is the class VectorActorView a single method, the implementation of the task drawRect:. In this method, we get a reference to the current graphics context by calling UIGraphicsGetCurrentContext and passing the context and the actor to the delegate to be drawn in the CGRect specified. Let's now take a look at a specific example.

Drawing a HealthBar

As mentioned, the vector-based actors used in this example are their own delegates in terms of drawing, so let's look at the HealthBar class again and see how it draws itself. See Listing 7–11.

Listing 7–11. *HealthBar.m (drawActor:WithContext:InRect:)*

```
-(void)drawActor:(Actor*)anActor WithContext:(CGContextRef)context InRect:(CGRect)rect{
    CGContextClearRect(context,rect);

    float height = 10;

    CGRect backgroundArea = CGRectMake(0, self.radius-height/2, self.radius*2, height);
    [self.backgroundColor setFill];
    CGContextFillRect(context, backgroundArea);

    CGRect healthArea = CGRectMake(0, self.radius-height/2, self.radius*2*percent,
height);
    [self.color setFill];
    CGContextFillRect(context, healthArea);

}
```

In Listing 7–11, we start by calling CGContectClearRect. This erases any old content in the given rect, getting us ready to do our real drawing. To draw the HealthBar, all we have to do is draw two rectangles, one for the background and another on top of the background representing the current percent of the health bar. To draw a rectangular area (as apposed to drawing the edges of a rectangle), we define a rect by calling CGRectMake. Then we specify the color to be used by calling setFill on the UIColor object backgroundColor, which is a property of HealthBar.

To actually change the color of the pixels of the UIView being drawn, we call CGContextFillRect passing in our context and the CGRect backgroundArea. To draw the foreground rectangle, we repeat the process, but the width of our top rectangle, healthArea, is calculated based the value of the property percent. This indicates that every time the percent of the HealthBar changes, we have to redraw the HealthBar. To make sure this happens, we have to provide our own implementation of setPercent: instead of relying on the default implementation that would be provided by a synthesized property. Listing 7–12 shows the custom implementation of setPercent: for the class HealthBar.

Listing 7–12. *HealthBar.m (setPercent:)*

```
-(void)setPercent:(float)aPercent{
    if (percent != aPercent){
        percent = aPercent;
        if (percent < 0){
            percent = 0;
        }
        if (percent > 1){
            percent = 1;
        }
        [self setNeedsViewUpdated:YES];
    }
}
```

Listing 7–12 shows our custom implementation of setPercent:. First we check to see if the values of percent (the backing variable for the property) and the passed in value aPercent have different values. If they do, we set percent to aPercent and we normalize the values by insuring it is between the values 0 and 1. Lastly, we call setNeedsViewUpdate: with Yes. This insures that the HealthBar will be redrawn during the next step in the animation. Let's now take a look at how the Bullet class is drawn.

Drawing the Bullet Class

We have looked at the class HealthBar and inspected how it is drawn. We only used a single drawing function, CGContextFillRect, to create our visual effect. Of course there are many, many more drawing functions that allow you to create any visual effect you want. We are not going to cover all of these functions, but there is fantastic documentation available online. That being said, let's take a look at the drawing implementation of the class Bullet. See Listing 7–13.

Listing 7–13. *Bullet.m (drawRect:WithContext:InRect:)*

```
-(void)drawActor:(Actor*)anActor WithContext:(CGContextRef)context InRect:(CGRect)rect{

    CGContextAddEllipseInRect(context, rect);
    CGContextClip(context);

    CGFloat locations[2];

    locations[0] = 0.0;
    locations[1] = 1.0;

    CGColorSpaceRef space = CGColorSpaceCreateDeviceRGB();

    UIColor* color1 = nil;
    UIColor* color2 = nil;

    if (damage >= 30){
        color1 = [UIColor colorWithRed:1.0 green:0.8 blue:0.8 alpha:1.0];
        color2 = [UIColor colorWithRed:1.0 green:0.0 blue:0.0 alpha:1.0];
    } else if (damage >= 20){
```

```
        color1 = [UIColor colorWithRed:0.8 green:1.0 blue:0.8 alpha:1.0];
        color2 = [UIColor colorWithRed:0.0 green:1.0 blue:0.0 alpha:1.0];
    } else {
        color1 = [UIColor colorWithRed:0.8 green:0.8 blue:1.0 alpha:1.0];
        color2 = [UIColor colorWithRed:0.0 green:0.0 blue:1.0 alpha:1.0];
    }

    CGColorRef clr[] = { [color1 CGColor], [color2 CGColor] };
    CFArrayRef colors = CFArrayCreate(NULL, (const void**)clr, sizeof(clr) /
sizeof(CGColorRef), &kCFTypeArrayCallBacks);


    CGGradientRef grad = CGGradientCreateWithColors(space, colors, locations);
    CGColorSpaceRelease(space);

    CGContextDrawLinearGradient(context, grad, rect.origin, CGPointMake(rect.origin.x +
rect.size.width, rect.origin.y + rect.size.height), 0);

    CGGradientRelease(grad);
}
```

In Listing 7–13, we see the code that draws each bullet on the screen. This code draws a circle with a gradient based on the damage of the Bullet. In my opinion, this is a lot of code to accomplish something so simple. Let's break it down piece by piece. The first line of the task creates an oval clipping region for the graphic context based on the provided rectangle. This means that future drawing operations will be visible only with this oval region.

The next step is to create the gradient used by the drawing operation. A gradient is made of two main parts. The first part is an array specifying where each color of the gradient should be applied. In our case, we are only using a two-color gradient, so we create an array called locations, making the first value 0 and second 1. This means that the colors should blend evenly from the start of the gradient to the end.

The second part of a gradient is the colors used. In our example, we specify each color based on value of damage. We simply assign the variables color1 and color2 to the colors we want to use to draw the Bullet. Once we have our colors, we create a CFArrayRef to store our colors in.

Once we have the locations of the colors and the colors themselves stored in the correct type of array, we call CGGradientCreateWithColors and pass the arrays in to create a CGGradientRef call grad. Note we also have passed in a reference to the color space, as defined by CGColorSpaceRef call ref. This is the default color space for the display and will only have to be changed if your application is doing sophisticated color presentation.

Once the CGGradientRef is created, call CGContextDrawLinearGradient. It would normally fill the entire UIView with the specified gradient, but because we specified the clip region at the top of the task, we get a nice circle with the gradient evenly drawn over it.

We have looked at how we can create actors that are drawn using Core Graphics to open up the possibility of dynamically drawing our actors, preventing us from having to create an image for every possible state of our actors. In the next section, we will look at another strategy for drawing actors, those based on particles.

Adding Particle Systems to Your Game

Particles systems in games are collections of many, usually small, images drawn together to create an overall affect. Particle systems are used in many different ways in computer graphics; some common examples are fire, water, hair, grass, and many more. In our example, we are going to create a new actor called Particle that will be used to draw out Comet actor and improve our Asteroid actor by making it appear to break up. Figure 7–3 shows this next example in action.

Figure 7–3. *Comets and destructible asteroids*

In Figure 7–3, we see three comets traveling across the screen. We also see a number of asteroids in various states of exploding. The asteroids drift in from off the screen, and when you tap the screen they explode into a number of smaller asteroids. In addition to creating the new, smaller asteroids, we also send out a number of particles that look like asteroids. This gives the destruction of the larger asteroid some added realism, because big chunks of rock seldom break cleanly into smaller pieces. These two different particle effects, the comets and the asteroid debris, are very different, but each is implemented with the same base class, Particle. Let's take a closer look at the class Example03Controller to better understand how this example works before diving into the details of each new actor. Listing 7–14, shows the important parts of Example03Controller.

Listing 7–14. *Example03Controller.m (doSetup and updateScene)*

```
-(BOOL)doSetup{
    if ([super doSetup]){
        NSMutableArray* classes = [NSMutableArray new];
        [classes addObject:[Asteroid class]];

        [self setSortedActorClasses:classes];

        UITapGestureRecognizer* tapRecognizer = [[UITapGestureRecognizer alloc]
initWithTarget:self action:@selector(tapGesture:)];
        [tapRecognizer setNumberOfTapsRequired:1];
        [tapRecognizer setNumberOfTouchesRequired:1];

        [actorsView addGestureRecognizer:tapRecognizer];

        return YES;
    }
    return NO;
}

-(void)updateScene{
    if (self.stepNumber % (60*5) == 0){
        [self addActor:[Comet comet:self]];
    }

    if ([[self actorsOfType:[Asteroid class]] count] == 0){
        int count = arc4random()%4+1;
        for (int i=0;i<count;i++){
            [self addActor:[Asteroid asteroid:self]];
        }

    }

    [super updateScene];
}
```

In Listing 7–14, we see two tasks of the class Example03Controller: doSetup and updateScene. The task doSetup follows our now familiar pattern of identifying the actor classes we want efficient access to, in this case Asteroid, by calling the task setSortedActorClass:. We also create a UITapGestureRecognizer called tapRecognizer and register that with the actorsView object. In this way, the task tapGesture: will be called when we tap the screen.

The task updateScene, as shown in Listing 7–14, adds a new Comet actor every five seconds (60*5) and adds one to four Asteroids when ever there are zero Asteroids in the scene. As with the previous examples, the code in the controller class is relatively simple. Most of the real logic for the example lives in the different actor classes. Before we look at the Asteroid class, let's consider the task tapGesture:, shown in Listing 7–15.

Listing 7–15. *Example03Controller.m (tapGesture:)*

```
- (void)tapGesture:(UIGestureRecognizer *)gestureRecognizer{
    for (Asteroid* asteroid in [self actorsOfType:[Asteroid class]]){
        [asteroid doHit:self];
    }
}
```

In Listing 7–15, we see that the `tapGesture:` task, which is called when the user taps the screen. It simply iterates through each `Asteroid` in the scene and calls doHit: on it. It is because of this task that we wanted to make sure we had easy access to the `Asteroids` in the scene.

The following section describes the implementation details of the actor `Asteroid`, as well the particle effect that is created when each `Asteroid` explodes.

Simple Particle System

The simplest particle systems involve creating a bunch of short-lived items on the screen that decay or disappear relatively quickly. Let's take a look at the `Asteroid` class and see how we can create a very simple particle effect.

The Asteroid Class

Before we look at how we add the particle effect to the destruction of the `Asteroid`, let's take a moment to understand the `Asteroid` class as whole. This will give us the context to understand the particle effect while building another `Actor` class. The header for the Asteroid class is shown in Listing 7–16.

Listing 7–16. *Asteroid.h*

```
@interface Asteroid : Actor{

}
@property (nonatomic) int level;
+(id)asteroid:(GameController*)aController;
+(id)asteroidOfLevel:(int)aLevel At:(CGPoint)aCenter;
-(void)doHit:(GameController*)controller;
@end
```

In Listing 7–16, we see the declaration of the class Asteroid. We see that `Asteroid` class extends `Actor` and has two constructors. The constructor `asteroid:` is used by the `Example03Controller` to add the biggest `Asteroids` into the scene. The constructor `asteroidOfLevel:At:` is used to create and add the smaller `Asteroids` in the task doHit:, which is called after a tap event. Listing 7–17 shows the first constructors for the class `Asteroid`.

Listing 7–17. *Asteroid.m (asteroid:)*

```
+(id)asteroid:(GameController*)acontroller{
    CGSize gameSize = [acontroller gameAreaSize];

    CGPoint gameCenter = CGPointMake(gameSize.width/2.0, gameSize.height/2.0);

    float directionOffScreen = arc4random()%100/100.0 * M_PI*2;
    float distanceFromCenter = MAX(gameCenter.x,gameCenter.y) * 1.2;

    CGPoint center = CGPointMake(gameCenter.x +
cosf(directionOffScreen)*distanceFromCenter, gameCenter.y +
sinf(directionOffScreen)*distanceFromCenter);

    return [Asteroid asteroidOfLevel:4 At:center];
}
```

In Listing 7–17, we see the task `asteroid:`, which creates a new `Asteroid` of size 4 by calling the other constructor. Most of the work being done is simply finding the starting point for the `Asteroid` off screen. This is identical to the code used to find a point for the `Powerup` actor in Chapter 6. The details can be found in Figure 6-3. The second constructor for `Asteroid` is shown in Listing 7–18.

Listing 7–18. *Asteroid.m (asteroidOfLevel:At:)*

```
+(id)asteroidOfLevel:(int)aLevel At:(CGPoint)aCenter{

    ImageRepresentation* rep = [ImageRepresentation
imageRepWithDelegate:[AsteroidRepresentationDelegate instance]];
    [rep setBackwards:arc4random()%2 == 0];
    if (aLevel >= 4){
        [rep setStepsPerFrame:arc4random()%2+2];
    } else {
        [rep setStepsPerFrame:arc4random()%4+1];
    }

    Asteroid* asteroid = [[Asteroid alloc] initAt:aCenter WithRadius:4 + aLevel*7
AndRepresentation:rep];

    [asteroid setLevel:aLevel];
    [asteroid setVariant:arc4random()%AST_VARIATION_COUNT];
    [asteroid setRotation: (arc4random()%100)/100.0*M_PI*2];

    float direction = arc4random()%100/100.0 * M_PI*2;
    LinearMotion* motion = [LinearMotion linearMotionInDirection:direction AtSpeed:1];
    [motion setWrap:YES];
    [asteroid addBehavior:motion];

    return asteroid;
}
```

In Listing 7–18, we see the main constructor of the class `Asteroid`. We also see that we create the `Asteroid` with a radius based on the value of `aLevel`. In this way, we have progressively smaller asteroids as they break apart, since each new asteroid has a level of one less than its creator. The last thing we do in the `asteroidOfLevel:At:` is to add a `LinearMotion` behavior the `Asteroid` so it moves in a straight line, wrapping around the screen.

Another thing to note in Listing 7–8 is that we are creating an ImageRepresentation in a slightly different way that we have up until this point. Previously, all the ImageRepresentations we created used the actor as the delegate, this made sense because it put all of the information about an Actor into a single file. However, we will want the Asteroid class and the Particles we create to look the same, though different size. To facilitate this, we have created a new class called AsteroidRepresentationDelegate that is responsible for specifying how things that look like asteroids are rendered. Let's move on to the class AsteroidRepresentationDelegate.

Representing an Asteroid and Particle with the Same Class

As mentioned, the definition of how an Asteroid is drawn in defined in the class AsteroidRepresentationDelegate. This class will be used to represent each Asteroid as well as the Particles we create when an Asteroid breaks up. The implementation of AsteroidRepresentationDelegate is shown in Listing 7–19.

Listiing 7–19. *AsteroidRepresentationDelegate.m*

```
+(AsteroidRepresentationDelegate*)instance{
    static AsteroidRepresentationDelegate* instance;
    @synchronized(self) {
                if(!instance) {
                        instance = [AsteroidRepresentationDelegate new];
                }
        }
        return instance;
}
-(int)getFrameCountForVariant:(int)aVariant AndState:(int)aState{
    return 31;
}
-(NSString*)getNameForVariant:(int)aVariant{
    if (aVariant == VARIATION_A){
        return @"A";
    } else if (aVariant == VARIATION_B){
        return @"B";
    } else if (aVariant == VARIATION_C){
        return @"C";
    } else {
        return nil;
    }

}
-(NSString*)baseImageName{
    return @"Asteroid";
}
```

In Listing 7–19, we see the expected tasks for defining an image-based actor. We see that image files will start with the string "Asteroid" as shown in the task baseImageName. We also see that there are 31 images for each of the three variants, as shown in getFramesCountForVariant:AndState: and getNameForVariant:.

The task, instance in Listing 7–19, is something new for us. This task represents a way for use to create a singleton of the class AsteroidRepresentationDelegate. We do this by synchronizing on the class object for Asteroid and only creating a new AsteroidRepresentationDelegate if the variable instance is nil. This allows us to only create one instance of this class, even though we could have hundreds of Asteroids and Particles using it to specify how they look. Let's now put all the pieces together and look at how an Asteroid is destroyed.

Destroying an Asteroid

We have set the groundwork for understanding how an Asteroid is created and represented on the screen. Let's take a look at the task doHit: to understand how these pieces are brought together to create the desired behavior. See Listing 7–20.

Listing 7–20. *Asteroid.m (doHit:)*

```
-(void)doHit:(GameController*)controller{
    if (level > 1){
        int count = arc4random()%3+1;
        for (int i=0;i<count;i++){
            Asteroid* newAst = [Asteroid asteroidOfLevel:level-1 At:self.center];
            [controller addActor:newAst];
        }
    }

    int particles = arc4random()%5+1;
    for (int i=0;i<particles;i++){
        ImageRepresentation* rep = [ImageRepresentation
imageRepWithDelegate:[AsteroidRepresentationDelegate instance]];
        Particle* particle = [Particle particleAt:self.center WithRep:rep Steps:25];
        [particle setRadius:6];
        [particle setVariant:arc4random()%AST_VARIATION_COUNT];
        [particle setRotation: (arc4random()%100)/100.0*M_PI*2];

        LinearMotion* motion = [LinearMotion linearMotionRandomDirectionAndSpeed];
        [particle addBehavior:motion];

        [controller addActor: particle];
    }
    [controller removeActor:self];
}
```

In Listing 7–20, we see the task doHit:. This task is called when we tap on the screen, causing all Asteroids in the scene to break apart into a number of smaller Asteroids. We also want to generate a number of asteroid-looking Particles when we do this. The first thing we do is make sure the Asteroid we call doHit: on is bigger than level one, because those asteroids should not create additional Asteroids. If the Asteroid is bigger than level one, we create one to three new Asteroids at the same location as the current Asteroid with one level smaller. These new Asteroids will travel away from the current Asteroid in a straight line until they also explode.

To create the particle effect in Listing 7–20, we first decide how many particles we will be adding to the scene. We then create a new Particle using the Asteroids current position, the singleton instance of AsteroidRepresentationDelegate, and indicate that the particle should live for 25 steps of animation. The radius of particle is set to 6. This is a purely aesthetic choice; it could be any value. The variation and rotation properties are also set to provide visual variation in each particle that is created. The last thing we do to the particle is specify a random `LinearMotion` before adding it to the scene. Figure 7–4 shows a close-up of the new `Asteroids` being created, as well as the Particles.

Figure 7–4. *Asteroids and particles, zoomed in*

The Particle Class

We looked at how the `Particles` are created when the each `Asteroid` explodes, so let's now look at the `Particle` class in detail and see how it is implemented, starting with the header file shown in Listing 7–21.

Listing 7–21. *Particle.h*

```
@interface Particle : Actor<ExpireAfterTimeDelegate> {

}
@property (nonatomic) float totalStepsAlive;
+(id)particleAt:(CGPoint)aCenter WithRep:(NSObject<Representation>*)rep
Steps:(float)numStepsToLive;
@end
```

In Listing 7–21, the header for the class Particle extends Actor and conforms to the protocol ExpireAfterTimeDelegate. `Particle` conforms to ExpireAfterTimeDelegate, because we want to change the opacity of the particle, fading it out, during the lifetime of the Particle. The property totalStepsAlive indicates how long the Particle should exists in the scene, and is set as part of the constructor particleAt:WithRep:Steps:. We can also see that a Particle requires a Representation to be passed as an argument to the constructor. The implementation of Particle is shown in Listing 7–22.

Listing 7–22. *Particle.m*

```
@implementation Particle
@synthesize totalStepsAlive;

+(id)particleAt:(CGPoint)aCenter WithRep:(NSObject<Representation>*)rep
Steps:(float)numStepsToLive{
    Particle* particle = [[Particle alloc] initAt:aCenter WithRadius:32
AndRepresentation:rep];
```

```
    [particle setTotalStepsAlive:numStepsToLive];

    ExpireAfterTime* expire = [ExpireAfterTime expireAfter:numStepsToLive];
    [expire setDelegate: particle];
    [particle addBehavior:expire];

    return particle;
}
-(void)stepsUpdated:(ExpireAfterTime*)expire In:(GameController*)controller{
    self.alpha = [expire stepsRemaining]/totalStepsAlive;
}
@end
```

In Listing 7–22, we see that the constructor particleAt:WithRep:Steps: is pretty straightforward. We create a new Particle object, passing in the provided Representation object. We also set the property totalStepsAlive and create an ExpiresAfterTime Behavior. Notice that particle is set as expire's delegate, this causes the task stepsUpdated:In: to be called every time expires is executed.

In the task stepsUpdated:In: we simply adjust the alpha value of the particle based on how far along the Particles life cycle we are. In a more complex implementation of a Particle, this fading behavior would be configurable, not just on or off, but also at which rate the particle fades. In this simple implementation, each particle simply fades in a linear way.

We have looked at the implementation of Asteroid and Particle and see that it is pretty simple to create a particle effect in our scene. In the next section we will look at the Comet actor and see a more visually striking example, though the implementation will be just as simple.

Creating Based Vector-Based Particles

In the previous example, we looked at Particles that were represented by a sequence of images. Because we have a flexible way of rendering actors, we can just as easily create particles that are programmatically drawn. In this example, we will be creating Particles that stay in one place, but we will move the point where they are created. This will give our Comet actors a nice glowing tail, like any comet should have. Figure 7–5 shows our new actor, the Comet.

Figure 7–5. *Comet detail*

In Figure 7–5 we see a zoomed in detail of a comet. It is composed of many particles, each one slowly fading in place to produce the tail of the comet. Let's take a look at the Comet class and see how this visual is created. Listing 7–23 shows the header for the Comet class.

Listing 7–23. *Comet.h*

```
enum{
    VARIATION_RED = 0,
    VARIATION_GREEN,
    VARIATION_BLUE,
    VARIATION_CYAN,
    VARIATION_MAGENTA,
    VARIATION_YELLOW,
    VARIATION_COUNT
};

@interface Comet : Actor<VectorRepresentationDelegate> {

}
+(id)comet:(GameController*)controller;
@end
```

In Listing 7–23, we see that Comet extends Actor and conforms to the protocol VectorRepresentationDelegate. There is a single constructor, taking a GameController as the only argument. Let's look at the implementation of this constructor, shown in Listing 7–24.

Listing 7–24. *Comet.m (comet:)*

```
+(id)comet:(GameController*)controller{
    CGSize gameSize = [controller gameAreaSize];

    CGPoint gameCenter = CGPointMake(gameSize.width/2.0, gameSize.height/2.0);

    float directionOffScreen = arc4random()%100/100.0 * M_PI*2;
    float distanceFromCenter = MAX(gameCenter.x,gameCenter.y) * 1.2;

    CGPoint center = CGPointMake(gameCenter.x +
cosf(directionOffScreen)*distanceFromCenter, gameCenter.y +
sinf(directionOffScreen)*distanceFromCenter);

    ImageRepresentation* rep = [ImageRepresentation imageRep];
    Comet* comet = [[Comet alloc] initAt:center WithRadius:16 AndRepresentation:rep];
    [rep setDelegate:comet];
    [comet setVariant:arc4random()%VARIATION_COUNT];

    float direction = arc4random()%100/100.0 * M_PI*2;
    LinearMotion* motion = [LinearMotion linearMotionInDirection:direction AtSpeed:1];
    [motion setWrap:YES];
    [comet addBehavior: motion];

    ExpireAfterTime* expire = [ExpireAfterTime expireAfter:60*15];
    [comet addBehavior: expire];

    return comet;
}
```

In Listing 7–24, we see that we again create each Comet off screen by calculating CGPoint called center that is on a circle just off the edge of the game area. We also create a VectorRepesentation and use the Comet as the delegate. We will be drawing the Comet and the Comet's tail particles with the same tasks. After specifying an ExpireAfterTime and a LinearMotion as the comet's behavior, it is ready to be added to the scene. Listing 7–25 shows the task that draws each Comet and Particle.

Listing 7–25. *Comet.m (drawActor:WithContect:InRect:)*

```objc
-(void)drawActor:(Actor*)anActor WithContext:(CGContextRef)context InRect:(CGRect)rect{
    CGContextClearRect(context,rect);
    CGColorSpaceRef space = CGColorSpaceCreateDeviceRGB();

    CGFloat locations[4];
    locations[0] = 0.0;
    locations[1] = 0.1;
    locations[2] = 0.2;
    locations[3] = 1.0;


    UIColor* color1 = nil;
    UIColor* color2 = nil;
    UIColor* color3 = nil;

    float whiter = 0.6;
    float c2alpha = 0.5;
    float c3aplha = 0.3;

    if (self.variant == VARIATION_RED){
        color1 = [UIColor colorWithRed:1.0 green:whiter blue:whiter alpha:1.0];
        color2 = [UIColor colorWithRed:1.0 green:0.0 blue:0.0 alpha:c2alpha];
        color3 = [UIColor colorWithRed:1.0 green:0.0 blue:0.0 alpha:c3aplha];
    } else if (self.variant == VARIATION_GREEN){
        color1 = [UIColor colorWithRed:whiter green:1.0 blue:whiter alpha:1.0];
        color2 = [UIColor colorWithRed:0.0 green:1.0 blue:0.0 alpha:c2alpha];
        color3 = [UIColor colorWithRed:0.0 green:1.0 blue:0.0 alpha:c3aplha];
    } else if (self.variant == VARIATION_BLUE){
        color1 = [UIColor colorWithRed:whiter green:whiter blue:1.0 alpha:1.0];
        color2 = [UIColor colorWithRed:0.0 green:0.0 blue:1.0 alpha:c2alpha];
        color3 = [UIColor colorWithRed:0.0 green:0.0 blue:1.0 alpha:c3aplha];
    } else if (self.variant == VARIATION_CYAN){
        color1 = [UIColor colorWithRed:whiter green:1.0 blue:1.0 alpha:1.0];
        color2 = [UIColor colorWithRed:0.0 green:1.0 blue:1.0 alpha:c2alpha];
        color3 = [UIColor colorWithRed:0.0 green:1.0 blue:1.0 alpha:c3aplha];
    } else if (self.variant == VARIATION_MAGENTA){
        color1 = [UIColor colorWithRed:1.0 green:whiter blue:1.0 alpha:1.0];
        color2 = [UIColor colorWithRed:1.0 green:0.0 blue:1.0 alpha:c2alpha];
        color3 = [UIColor colorWithRed:1.0 green:0.0 blue:1.0 alpha:c3aplha];
    } else if (self.variant == VARIATION_YELLOW){
        color1 = [UIColor colorWithRed:1.0 green:1.0 blue:whiter alpha:1.0];
        color2 = [UIColor colorWithRed:1.0 green:1.0 blue:0.0 alpha:c2alpha];
        color3 = [UIColor colorWithRed:1.0 green:1.0 blue:0.0 alpha:c3aplha];
    }
    UIColor* color4 = [UIColor colorWithRed:0.0 green:0.0 blue:0.0 alpha:0.0];
```

```
    CGColorRef clr[] = { [color1 CGColor], [color2 CGColor] , [color3 CGColor], [color4
CGColor]};
    CFArrayRef colors = CFArrayCreate(NULL, (const void**)clr, sizeof(clr) /
sizeof(CGColorRef), &kCFTypeArrayCallBacks);

    CGGradientRef grad = CGGradientCreateWithColors(space, colors, locations);
    CGColorSpaceRelease(space);

    CGContextDrawRadialGradient(context, grad, CGPointMake(self.radius, self.radius), 0,
CGPointMake(self.radius, self.radius), self.radius, 0);

    CGGradientRelease(grad);
}
```

In Listing 7–25, we want to draw a simple radial gradient with the color appropriate to the variant at the center and transparent at the edge. We have to create a four-color gradient to achieve a nice looking comet. The first color is mostly white and fully opaque, the next two colors are the color of the variant with two different alpha values. The last color will be completely transparent. The functions CGContextDrawRadialGradient draws the radial gradient at the center of the actor with the first color and create a smooth transition to the transparent edge of the actor.

Now that we can see how each particle is drawn, we should look how the particles are created, Listing 7–26 shows the implementation of Comet's step: task.

Listing 7–26. *Comet.m (step:)*

```
-(void)step:(GameController*)controller{
    if ([controller stepNumber]%3 == 0){
        VectorRepresentation* rep = [VectorRepresentation
vectorRepresentation];
        [rep setDelegate:self];

        int totalStepsAlive = arc4random()%60 + 60;
        Particle* particle = [Particle particleAt:self.center WithRep:rep
Steps:totalStepsAlive];
        [particle setRadius:self.radius];
        [controller addActor: particle];
    }
}
```

In Listing 7–26, the step: task creates a new Particle every three steps in the animation. For each new Particle, a VectorRepresentation is created, but note that the delegate is the Comet object. The Particle is created with a random lifespan from 1 second to 2 seconds. This randomness creates the slight flickering variation in the tail of the comet. Notice that the Particle has no additional behaviors added at this point, this causes the Particle to just sit in one place and fade away.

Summary

In this chapter we looked at a technique to create actors in our game that are drawn programmatically. We created a new class call VectorRepresentation to handle the details of create a UIView for our code to draw on. We used these new vector-based actors to create a health bar and a simple bullet. The chapter continued by showing two simple examples of creating particle systems. The first example reused the art for the asteroids to create a sense of debris when we exploded the asteroids. The second example used vector-based particles to create acomets with glowing tails. The actors created in this and the previous chapter will be used to create a complete game in Chapter 10.

Chapter 8

Building Your Game: Understanding Gestures and Movements

The really neat thing about creating games for mobile devices is the unique way that the user interacts with the game. The complex gestures supported by iOS are a relatively new way to interact with a computer, and new interface techniques can give users trouble if they are unfamiliar with them. However, users of iOS are already familiar with the concepts of double taps, pinches, and swipes, so including these gestures in your game does not present a usability problem.

In this chapter, we will explore how you can add gestures to your application or game using the built-in libraries provided by iOS. We will also explore how to attach specific gestures to specific actors in a game. By the end of this chapter, you will know how to work with all supported gesture and be able to use them to drive user interaction in a game. We will be using the familiar classes of GameController, Actor, and the rest in our examples. So your understanding of these gesture events will be tied directly into what you have learned in the previous chapters.

Beyond simply touching the screen, users can interact with iOS 5 applications through the accelerometer or the gyroscope included in the newer devices. Using the motion of the device as input to a game opens up a new dimension to the user experience. We will explore how to use the sensors and tie them into game events.

Touch Input: The Basics

There are really three ways to accept touch input from the user. The first is to use buttons or other prebuilt widgets and implement tasks that respond to the user events. I am not going to go through the details of this type of input, as it is the most basic, and is well covered in all iOS introductory books. The second method is to create subclasses of UIView and implement a number of touch-related tasks. The last technique is to wire up one of the six UIGestureRecognizer classes to a UIView and register an object to respond to the different gestures.

I think the last technique is the most robust in most cases, because it provides a way for users to interact with your application using a set of known gestures. However, there are perfectly good reasons to use the second technique, implementing the touch related tasks, especially if the user interaction you are shooting for is not a gesture per se. An example might be a drawing application, where you simply want access to all points of contact on the screen to do your drawing.

The following section will review the touch-related tasks of UIView. We will then move on and explore UIGestureRecognizer classes and motion events.

The sample code that accompanies this chapter is found in the project Sample 08. While most of the code in this book can be run on either a device or the emulator, I highly recommend running this code on an iPhone (or iPod touch)—if possible, iOS 4.2 or later—because the complex gestures are hard to simulate on the emulator. To really get a feel for the user experience, it is best it perform the actual gestures with your hand and see how each one feels.

Extending UIView to Receive Touch Events

UIView has four tasks that get called when a user touches it on the screen. I am collectively calling these the touch-related tasks. They are shown in Listing 8–1.

Listing 8–1. *The four touch related tasks*

```
-(void)touchesBegan:withEvent:
-(void)touchesCancelled:withEvent:
-(void)touchesEnded:withEvent:
-(void)touchesMoved:withEvent:
```

The first argument of each of these tasks is an NSSet that contains a UITouch for each discrete touch on the screen. The second argument is a UIEvent that is basically a generic wrapper for the three different input types available on iOS: touch events, motion events, and remote-control events. We won't be covering remote-control events; those are events generated by accessories, like an external keyboard.

The task touchesBegan:withEvent: is called as soon as a finger (or fingers) touch the screen. The task touchesEnded:withEvent: is called when the fingers are removed. If the fingers slide around, the touchesMoved:withEvent: would be called. The task touchesCancelled:withEvent: is call if the system has to take control of the screen. For example, if another application is launched, the touchesCancelled:withEvent: is called on the first application before control is passed to the new app. The sample code provides an example of using these tasks, as shown in Figure 8–1.

Figure 8–1. *Touch events*

In Figure 8–1, we see a several lines comprised of dots. Each one of these lines is the result of a finger being dragged across the screen. The top-most line was created when a single figure was moved across the screen—it is drawn with red dots. The second top-most line was created by dragging two fingers across the screen, as was the one below it, and both are drawn with green dots. The bottom three lines were drawn with three fingers and is made of blue dots. If you can see Figure 8–1 in color (if you have the ebook), take a look at the left-most dots on the bottom line. Notice that one is red and the other is green. This is because I did not get all three of my fingers to land on the screen at the same time.

In this first example, besides drawing dots with your finger, you can scale the dots up or down by tapping with a single finger or two fingers respectively. The scaling feature demonstrates how touch events and gestures can work together. We will explore this in more detail once we understand the basics of the touch events.

Looking At the Event Code

The implementation of this simple demonstration shows how we receive the touch events and create actors out of each touch on the screen. There are two classes involved: TouchEventsController that extends GameController and TouchEventsView that extends UIView. Let's start by looking the class TouchEventsView. The header is shown in Listing 8–2.

Listing 8–2. *TouchEventsView.h*

```
@interface TouchEventsView : UIView{
    IBOutlet TouchEventsController* controller;
    NSMutableArray* sparks;
}

@end
```

In Listing 8–2, we see an IBOutlet so we can wire the TouchEventsView up to a TouchEventsController. We also see an NSMutableArray called sparks. This collection is used to keep track of sparks as they are being added; as mentioned, the touch event can be cancelled. By keeping track of the sparks that we add, we can clean up any sparks that are added before touchCancelled:withEvent: is called. Let's take a look at the implementation of the class TouchEventsView, as shown on Listing 8–3.

Listing 8–3. *TouchEventsView.m*

```
@implementation TouchEventsView

- (void)touchesBegan:(NSSet *)touches withEvent:(UIEvent *)event{
    NSLog(@"Begin: %i", [touches count]);
    sparks = [NSMutableArray new];
}
- (void)touchesCancelled:(NSSet *)touches withEvent:(UIEvent *)event{
    NSLog(@"Cancelled: %i", [touches count]);

    for (Spark* spark in sparks){
        [controller removeActor:spark];
    }

    [sparks removeAllObjects];
}
- (void)touchesEnded:(NSSet *)touches withEvent:(UIEvent *)event{
    NSLog(@"Ended: %i", [touches count]);
    int count = [touches count];

    for (UITouch* touch in touches){
        Spark* spark = [Spark spark:count-1 At:[touch locationInView:self]];
        [controller addActor:spark];
        [sparks addObject:spark];
    }

    [sparks removeAllObjects];
}
- (void)touchesMoved:(NSSet *)touches withEvent:(UIEvent *)event{
    NSLog(@"Moved: %i", [touches count]);
    int count = [touches count];

    for (UITouch* touch in touches){
        Spark* spark = [Spark spark:count-1 At:[touch locationInView:self]];
        [controller addActor:spark];
        [sparks addObject:spark];
    }
```

```
}
@end
```

In Listing 8–3, we see the four tasks discussed earlier. The class UIResponder defines each task, which is the super class of UIView. These methods are called as the user interacts with the screen. By implementing them, we can add our custom behavior. The task touchesBegan:withEvent: is called when a finger touches the screen. When this happens, we create a new NSMutableArray to keep track of all of the sparks we are going to add. Any of the other three tasks could be called next. If the user moves his or her finger, or places another new finger on the screen, the task touchesMoved:withEvent: is called. In this task, we create and a new Spark for each touch. We use the number of touches, minus 1, to specify the variant or color of the Spark, which is then added the scene. The spark is just a simple particle-like class that removes itself after five seconds.

There are two ways that a touch event can end. The first way is when touchesEnded:withEvent: is called after the user has removed all of his fingers. In this case, we do the same thing we did in touchesMoved:withEvent:: we create a Spark for each touch. We also clear the NSMutableArray sparks, to remove reference to the spark objects.

The other way a touch event can end is by it being cancelled. An event is cancelled when something else in the system happens that makes completing the event inappropriate. When this tasks is called, we removed all of the Spark objects we have created since touchesBegan:withEvent: was last called, undoing the work we have done up until this point.

The touchesCancelled:withEvent: task can be called when switching applications or having a dialog popup. Another perhaps more common way for touch events to be cancelled is if a gesture recognizer is triggered. A gesture recognizer, as defined by the class UIGestureRecognizer, is an object that inspects the touch events on a UIView and responds when the touches can be considered a specific gesture, like a tap, pinch, or swipe.

To experiment with the touchesCancelled:withEvent: tasks, do a two-finger double tap. Notice that after the first tap, there are two Sparks where you fingers touched. After the second tap, the Spark objects that where added get removed. This is because touchesCancelled:withEvent: was called when the two-figure double-tap gesture recognizer was activated by the touch events.

The fact that gesture recognizers are able to cancel touch events allows developers to mix and match any number of gestures together on a UIView without worrying about the gestures conflicting. In addition to canceling the touch events, gestures can also cancel each other. For example, a pinch gesture and a two-finger drag gesture both start the same way, with two fingers landing on the screen at about the same time. These two gestures could easily conflict if there was no mechanism in place to cancel each other. Let's move on a see how these touch events apply to the game related classes.

Applying Touch Events to Actors

Let's take a quick look at the class TouchEventsController and see how we set up the demo to respond to tap gestures. The important part of TouchEventsController.m is shown in Listing 8–4.

Listing 8–4. *TouchEventsController.m (doSetup)*

```
-(BOOL)doSetup{
    if ([super doSetup]){

        [self setGameAreaSize:CGSizeMake(320, 480)];
        [self setSortedActorClasses:[NSMutableArray arrayWithObject:[Spark class]]];

        UITapGestureRecognizer* doubleTapTwoTouch = [[UITapGestureRecognizer alloc]
initWithTarget:self action:@selector(doubleTap:)];
        [doubleTapTwoTouch setNumberOfTapsRequired:2];
        [doubleTapTwoTouch setNumberOfTouchesRequired:2];


        UITapGestureRecognizer* doubleTapOneTouch = [[UITapGestureRecognizer alloc]
initWithTarget:self action:@selector(doubleTap:)];
        [doubleTapOneTouch setNumberOfTapsRequired:2];
        [doubleTapOneTouch setNumberOfTouchesRequired:1];

        [actorsView addGestureRecognizer:doubleTapOneTouch];
        [actorsView addGestureRecognizer:doubleTapTwoTouch];

        return YES;
    }
    return NO;
}
```

In Listing 8–4, we see the task doSetup, from the class TouchEventsController. This task is called when this UIViewController instance's view is loaded. After setting up the game area size and marking the class Spark as a sorted actor class, we create two instances of UITapGestureRecognizer. The first instance of UITapGestureRecognizer is called doubleTapTwoTouch and is configured to respond when two fingers do a double tap. The second UITapGestureRecognizer is called doubleTapOneTouch and is configured for a single-finger double tap. Both of these UITapGestureRecognizer are added to actorsView with the task addGestureRecognizer:.

Looking at the constructor used to create these gesture recognizers, we see a selector referencing the task doubleTap:. This task is called whenever one of these two tasks recognizes a gesture. We could have had each gesture recognizer call different tasks, and that may be desirable in a more complex application. In this example, it is easy enough to distinguish these two gesture recognizers in the doubleTap: tasks, as shown in Listing 8–5.

Listing 8–5. *TouchEventsController.m (doubleTap:)*

```
-(void)doubleTap:(UITapGestureRecognizer*)doubleTap{
    float scale;
    if ([doubleTap numberOfTouches] == 1){
```

```
        scale = 2.0;
    } else {
        scale = 0.5;
    }
    NSLog(@"Touches: %i", [doubleTap numberOfTouches]);

    for (Spark* spark in [self actorsOfType:[Spark class]]){
        float radius = [spark radius]*scale;
        if (radius < 2){
            radius = 2;
        } else if (radius > 128){
            radius = 128;
        }
        [spark setRadius:radius];
    }
}
```

In Listing 8–5, we check to see if the UITapGestureRecognizer doubleTap that triggered this task is configured for a single touch. Based on this result of this check, we set the value of scale to either 2.0 or 0.5. In this way, a single-finger double tap will increase the scale of the Spark objects, and a two-finger double tap will reduce the size of the Spark objects. Once we have the correct value for scale, we simply iterate over all of the Spark objects in the scene and adjust their radius based on scale.

We have looked at touch events so we can understand the touch event lifecycle. We also took a quick look at the UITapGestureRecognizer class. The following section will discuss the UITapGestureRecognizer in more detail.

Understanding Gesture Recognizers

iOS allows the user to interact with an application with a rich set of gestures. These include taps, swipes, pinches, and so on. Each of these gestures is comprised of some reasonably complex logic to interpret touch events into a cohesive gesture. If this was left up to each developer to implement, the user experience with be pretty bad, as each developer would inevitably make different assumptions about each gesture. For example, the long press gesture has a default duration. We would want this consistent across all applications, because we don't want to "retrain" each user for each application as to what a long press is. In this spirit, Apple provides a number of classes that are designed to identify the most common gestures a developer is going need. Each of the gesture recognizers extends the abstract base class UIGestureRecognizer and is shown in Listing 8–6.

Listing 8–6. *Concrete Implementations of* UIGestureRecognizer

```
UITapGestureRecognizer
UIPinchGestureRecognizer
UIRotateGestureRecognizer
UISwipeGestureRecognizer
UIPanGestureRecognizer
UILongPressGestureRecognizer
```

In Listing 8–6, we see each subclass of UIGestureRecognizer that is built into iOS 5. Tap is probably the most common gesture iOS, and needs no description. The pinch gesture is when two fingers either move closer or farther away from each other. The rotation gesture is when two fingers move on the screen like you are opening a bottle of ketchup. The swipe gesture is when one or more fingers are flicked across the screen, as if turning a page in a magazine. The pan gesture is similar to swipe, but slower and more deliberate. The long press gesture is much like a tap, but the finger is held to the screen longer.

Each of these classes provides an easy way for a developer to request that a task be called should any of the gestures happen on a particular UIView. In addition to calling a task, each gesture happens over time and will be in one of several states as the callback task is called. These states are shown in Listing 8–7.

Listing 8–7. *Values for the enum UIGestureRecognizerState.*

```
UIGestureRecognizerStatePossible
UIGestureRecognizerStateBegan
UIGestureRecognizerStateChanged
UIGestureRecognizerStateEnded
UIGestureRecognizerStateCancelled
UIGestureRecognizerStateFailed
UIGestureRecognizerStateRecognized = UIGestureRecognizerStateEnded
```

In Listing 8–7, we see the seven possible states for a gesture recognizer to be in, and are analogous to four tasks that are called touch events (see Listing 8–1). The default state is UIGestureRecognizerStatePossible and is the state that all UIGestureRecignizer instances will be in if no touch events are happening. UIGestureRecognizerStateBegan is the state when a gesture has started, but not yet changed or completed. For a pinch gesture, this would be when both fingers have landed on the screen. UIGestureRecognizerStateChanged is the state of gesture in progress. A gesture will be over when one of the three following states has occurred, UIGestureRecognizerStateEnded, UIGestureRecognizerStateCancelled, or UIGestureRecognizerStateFailed. The state UIGestureRecognizerStateEnded corresponds the successful completion of a gesture.
UIGestureRecognizerStateCancelled is the state of a gesture that cannot be completed.
UIGestureRecognizerStateFailed is the state of a gesture that has received touch events in contradiction to the gesture. The last state, UIGestureRecognizerStateRecognized, has the same value as UIGestureRecognizerStateEnded and simply provides a semantic difference that might be useful in code.

While the states in Listing 8–7 provided a comprehensive description of the states a gesture recognizer can go through, not all gesture recognizers use all of these states. For example, UITapGestureRecognizer is known as a "discrete" gesture recognizer and does not report changes. Further, discrete gestures cannot fail or be cancelled. They simply call the selector task once, with the state of UIGestureRecognizerStateRecognized (e.g., UIGestureRecognizerStateEnded).

It is possible to implement a new UIGestureRecognizer, but we will not be covering that. We will, however, take each gesture recognizer and explore how it works with an example.

Tap Gestures

Tap gestures are used throughout iOS. From launching an application to dropping a pin in the Maps application, tap gestures are the backbone of the iOS interface. This makes sense, because they are very similar to mouse clicks on desktop operating systems. There are similarities between the tap gesture and the mouse click; for example, launching an application from the system dock in OS X is done with a single click, while launching an iOS application is done with a single tap. The analogy continues if we consider the "secondary click" (also known as a control click or right click) on OS X; this is similar to the two-finger tap in iOS. In fact, if you are using a Mac laptop, the touch pad can be configured to treat a two-finger tap as a "secondary click."

For this example, we are going to configure a bunch of UITapGestureRecognizers to respond to different combinations of tap counts and the number of fingers involved (touches). Figure 8–2 shows this example in action.

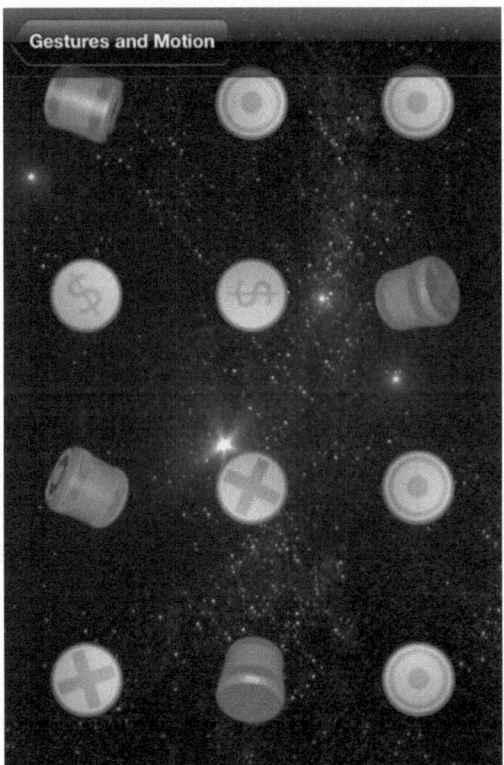

Figure 8–2. *Power-ups activated by tap gestures*

In Figure 8–2, we see 12 power-ups organized into three columns and four rows. The power-ups start disabled, like the one in the lower right. When the user taps the screen, a power-up enables, becoming brighter and spinning. The power-up that is enabled is based on the number of fingers used in the gesture and whether one, two, or three fingers are used. For example, a two-finger triple tap enabled the right-most power-up on the second row. Similarly, a three-finger single tap enabled the left-most power-up on the third row. So, in short, the columns show the number of taps and the rows the number fingers involved. If you want to waste an hour of your life, try and get all of the power-ups spinning. I *think* I got it to happen, but I could not get a screenshot in time to prove it.

This demo is implemented in the class TapGestureController. Listing 8–8 shows the file TapGestureController.h.

Listing 8–8. *TapGestureController.h*

```
@interface TapGesutureController : GameController<TemporaryBehaviorDelegate>{
    NSMutableArray* powerups;
}

- (void)tapGesture:(UITapGestureRecognizer *)sender;

@end
```

In Listing 8–8, we see the header for the class TapGestureController and that it extends GameController and conforms to the protocol TemporaryBehaviorDelegate. We also see that there is an NSMutableArray called powerups that is used to store the twelve power-ups in this example. In other examples, we have relied on GameController to keep track of different types of actors by specifying classes that should be sorted. In this example, we are using a separate NSMutableArray because we want to use the order of the Powerup objects in powerups to keep track of the row and column. Lastly, we see the declaration of the task tapGesture: that is called when a UITapGestureRecognizer recognizes a gesture. The implementation of the setup code for TapGestureController is shown in Listing 8–9.

Listing 8–9. *TapGestureController.m (doSetup)*

```
-(BOOL)doSetup{
    if ([super doSetup]){
        [self setGameAreaSize:CGSizeMake(320, 480)];
        powerups = [NSMutableArray new];

        for (int tap=0;tap<=2;tap++){
            for (int touch=0;touch<=3;touch++){
                float x = 320.0/6.0 + tap*320.0/3;
                float y = 480.0/8.0 + touch*480/4;
                CGPoint center = CGPointMake(x, y);

                Powerup* powerup = [Powerup powerup:self At:center];

                [self addActor: powerup];
                [powerups addObject:powerup];

            }
        }

        for (int touch=1;touch<=4;touch++){
            UITapGestureRecognizer* tripleTap = [[UITapGestureRecognizer alloc]
initWithTarget:self action:@selector(tapGesture:)];
```

```
            [tripleTap setNumberOfTapsRequired:3];
            [tripleTap setNumberOfTouchesRequired:touch];

            UITapGestureRecognizer* doubleTap = [[UITapGestureRecognizer alloc]
initWithTarget:self action:@selector(tapGesture:)];
            [doubleTap setNumberOfTapsRequired:2];
            [doubleTap setNumberOfTouchesRequired:touch];
            [doubleTap requireGestureRecognizerToFail:tripleTap];


            UITapGestureRecognizer* singleTap = [[UITapGestureRecognizer alloc]
initWithTarget:self action:@selector(tapGesture:)];
            [singleTap setNumberOfTapsRequired:1];
            [singleTap setNumberOfTouchesRequired:touch];
            [singleTap requireGestureRecognizerToFail:doubleTap];


            [actorsView addGestureRecognizer:tripleTap];
            [actorsView addGestureRecognizer:doubleTap];
            [actorsView addGestureRecognizer:singleTap];
        }

        return YES;
    }
    return NO;
}
```

In Listing 8–9, we see the task doSetup for the class TapGestureController. In this task, we generate the 12 Powerup objects and add them as actors. We also keep track of each Powerup in the NSMutableArray powerups. After creating the Powerup objects, we create 12 UITapGestureRecognizers—done for each Powerup. Each UITapGestureRecognizer is added to actorsView with the task addGestureRecognizer:. Note that we are creating UITapGestureRecognizer for each possible combination of taps and touches, instead of trying to use a single UITapGestureRecognizer to recognize the various combinations. Although it is possible in some cases to do this, everything works a lot better when you are as concise as possible with UIGestureRecognizers, as there are some subtle interactions between gesture recognizers. For example, the recognizer responsible for detecting triple taps has to cancel those responsible for double and single taps.

Each UITapGestureRecognizer is configured to call tapGesture: when a gesture is recognized, as shown in Listing 8–10.

Listing 8–10. *TapGestureController.m (tapGesture:)*

```
- (void)tapGesture:(UITapGestureRecognizer *)sender{
    int taps = [sender numberOfTapsRequired];
    int touches = [sender numberOfTouches];

    int index = (taps-1)*4+(touches-1);

    Powerup* powerup = [powerups objectAtIndex:index];

    TemporaryBehavior* tempBehav = [TemporaryBehavior temporaryBehavior:nil for:60*5];
    [tempBehav setDelegate:self];

    NSMutableArray* behaviors = [powerup behaviors];
```

```
    [behaviors removeAllObjects];
    [behaviors addObject:tempBehav];

    [powerup setState:STATE_GLOW];
}
```

In Listing 8–10, we see the task `tapGesture:`, which is called whenever one of the 12 `UITapGestureRecognizers` recognizes a tap gesture. Note that we are not checking the state of the `UITapGestureRecognizer`, because tap gestures are considered discrete and don't use the state information. We use the tasks `numberOfTapsRequired` and `numberOfTouches` to figure out which `UITapGestureRecognizer` has responded and store these values in taps and touches, respectively. These values tell us the index of the Powerup that should be enabled and is found in the `NSMutableArray` powerups. Once we have the correct Powerup, we create a `TemporaryBehavior` and set `self` as the delegate. The power-up then has its behavior set to the newly created `TemporaryBehavior`, as well as have its state set to `STATE_GLOW`. This starts the power-up spinning. Let's take a closer look at the class `TemporaryBehavior`.

The TemporaryBehavior Class

The class `TemporaryBehavior` applies a `Behavior` to an actor for a fixed number of steps. In this case, we don't really want to apply any behaviors; we just want to know when five seconds has gone by. Let's take a quick look the class `TemporaryBehavior` and see how this works. The declaration of `TemporaryBehavior` is shown in Listing 8–11.

Listing 8–11. *TemporaryBehavior.h*

```
@class TemporaryBehavior;
@protocol TemporaryBehaviorDelegate
-(void)stepsUpdatedOn:(Actor*)anActor By:(TemporaryBehavior*)tempBehavior
In:(GameController*)controller;
@end

@interface TemporaryBehavior : NSObject <Behavior>{

}
@property (nonatomic) long stepsRemaining;
@property (nonatomic, strong) NSObject<Behavior>* behavior;
@property (nonatomic, strong) NSObject<TemporaryBehaviorDelegate>* delegate;

+(id)temporaryBehavior:(NSObject<Behavior>*)aBehavior for:(long)aNumberOfSteps;

@end
```

In Listing 8–11, we see this definition of the class `TemporaryBehavior`. We see that it defines a protocol called `TemporaryBehaviorDelegate`, which `TapGestureController` conforms to (see Listing 8–8). The constructor for `TemporaryBehavior` takes a `Behavior` and a number of steps as arguments. Listing 8–12 shows the implementation of `TemporaryNehavior`.

Listing 8–12. *TemporaryBehavior.m*

```
+(id)temporaryBehavior:(NSObject<Behavior>*)aBehavior for:(long)aNumberOfSteps{
    TemporaryBehavior* temp = [TemporaryBehavior new];
    [temp setBehavior:aBehavior];
    [temp setStepsRemaining:aNumberOfSteps];

    return temp;
}
-(void)applyToActor:(Actor*)anActor In:(GameController*)gameController{
    stepsRemaining--;

    [behavior applyToActor:anActor In:gameController];
    [delegate stepsUpdatedOn:anActor By:self In:gameController];


    if (stepsRemaining <= 0){
        [[anActor behaviors] removeObject:self];
    }
}
```

`@end`

In Listing 8–12, we see the implementation of the class TemporaryBehavior. The constructor simply creates a TemporaryBehavior object and populates it with the provided arguments. The task, applyToActor:In:, comes from the protocol Behavior (which TemporaryBehavior conforms to), and is called for every step of the game. In this task, we simply apply the provided behavior to the actor and inform the delegate that this work has been done. Continuing with tap gesture example, let's look at how TapGestureController responds to the task stepsUpdatedOn:By:In:, as shown in Listing 8–13.

Listing 8–13. *TapGestureController.m (stepsUpdatedOn:By:In:)*

```
-(void)stepsUpdatedOn:(Actor*)anActor By:(TemporaryBehavior*)tempBehavior
In:(GameController*)controller{
    if ([tempBehavior stepsRemaining]==0){
        [anActor setState:STATE_NO_GLOW];
    }
}
```

In Listing 8–13, we see the task stepsUpdatedOn:By:In: as defined by the class TapGestureController. In this task, we simply see if the TemporaryBehavior, tempBehavior, is at the end of its life. If it is, we set the actor's state to STATE_NO_GLOW, which stops the power-up from spinning.

We have looked in detail at the different configurations available for UITapGestureRecognizer. The following section will explore a new one: the pinch.

Pinch Gestures

Pinch gestures show up in iOS usually as a way of zooming in and out or scaling something on the screen. This gesture is started by placing two fingers down and then by moving them either toward each other or away. Figure 8–3 shows this gesture in action.

Scaled Up   Scaled Down

Figure 8–3. *Pinch gesture changing the size of a saucer*

In Figure 8–3, we see our example of using pinch gestures. On the left, the saucer is scaled up as the result of a pinch-out gesture. On the right we see the same saucer scaled down as the result of a pinch-in gesture. As the saucer is being scaled, it stops spinning. We take advantage of the different states of the pinch gesture to stop and start the spinning.

A pinch gesture can be detected by adding an instance of UIPinchGestureRecognizer to a UIView. A UIPinchGestureRecognizer will not only inform you when a pinch gesture occurs, but also reports a scale and velocity value associated with the gesture. Having the scale reported by a UIPinchGestureRecognizer is very handy, because this is such a common use of the pinch gesture.

This example is implemented in the class PinchGestureController. Let's take a quick look at the header file, as shown in Listing 8–14.

Listing 8–14. *PinchGestureController.h*

```
@interface PinchGestureController : GameController{
    Saucer* saucer;
    float startRadius;
}
-(void)pinchGesture:(UIPinchGestureRecognizer*)sender;
@end
```

In Listing 8–14, we see the header file for the class PinchGestureController, like the previous example it extends GameController. It also maintains a reference to the Saucer object, as well the radius of the Saucer object at the beginning of a gesture. The task, pinchGesture: is called by the gesture recognizer. The implementation of this class will shed more light on how this example works, starting with the doSetup task, as shown in Listing 8–15.

Listing 8–15. *PinchGestureController (doSetup)*

```
-(BOOL)doSetup{
    if ([super doSetup]){
        [self setGameAreaSize:CGSizeMake(320, 480)];

        saucer = [Saucer saucer:self];
        [self addActor:saucer];

        UIPinchGestureRecognizer* pinchRecognizer = [[UIPinchGestureRecognizer alloc]
initWithTarget:self action:@selector(pinchGesture:)];
        [actorsView addGestureRecognizer:pinchRecognizer];

        return YES;
    }
    return NO;
}
```

In Listing 8–15, we have a very simple doSetup task for the class PinchGestureController. We simply add a new Saucer and setup a UIPinchGestureRecognizer to call pinchGesture: when a pinch gesture is recognized. Let's take a look at how we respond the gesture in the following section.

Responding to the Pinch Gesture

We have looked at how to register the pinch gesture recognizers. The implementation of pinchGesture: is shown in Listing 8–16.

Listing 8–16. *PinchGestureController (pinchGesture:)*

```
-(void)pinchGesture:(UIPinchGestureRecognizer*)pinchRecognizer{
    if (pinchRecognizer.state == UIGestureRecognizerStateBegan){
        startRadius = [saucer radius];
        [saucer setAnimationPaused:YES];
    } else if (pinchRecognizer.state == UIGestureRecognizerStateChanged){
        float scale = [pinchRecognizer scale];
        float velocity =  [pinchRecognizer velocity];

        NSLog(@"Scale: %f Velocty: %f",scale, velocity);
```

```
        float radius = startRadius*scale;
        if (radius < 8){
            radius =  8;
        } else if (radius > 128.0){
            radius = 128;
        }

        [saucer setRadius:radius];
    } else if (pinchRecognizer.state == UIGestureRecognizerStateEnded){
        [saucer setAnimationPaused:NO];
    } else if (pinchRecognizer.state == UIGestureRecognizerStateCancelled){
        [saucer setRadius:startRadius];
        [saucer setAnimationPaused:NO];
    }
}
```

In Listing 8–16, we see the task pinchGesture: that is called whenever a pinch gesture is detected on the UIView actorsView. In this example, we are checking the state of the gesture recognizer. The state information indicates if the gesture has just started (UIGestureRecognizerStateBegan), is in progress (UIGestureRecognizerStateChanged), or has terminated (UIGestureRecognizerStateEnded or UIGestureRecognizerStateCancelled).

When the gesture starts, we record the radius of saucer and store it in the variable startRadius. We also call setAnimationPaused: on saucer to stop if from animating while being scaled. It is not a requirement to stop the animation. We are simply doing this so we can observe the change of the gesture recognizer's state as we play with the sample on our phone.

When the gesture is in progress, we get the scale and velocity of the gesture. The scale value is the change in scale since the beginning of the gesture. To put it another way, the scale is accumulative, and that's why we apply the scale to the startRadius and not to the current radius of saucer. Once we calculate our new radius, we apply that to saucer. The velocity is the scale factor per second. This value can be used to calculate how much drift to apply to an animation after the gesture is over. We are not using the value except to log it.

When the gesture has ended, normally we simply set the animationPaused property of saucer to NO. This starts the saucer animating again. If the gesture was cancelled, we reset the saucer back to its original radius and un-pause the animation. To test the cancel state, start with the saucer scaled all the way down, start scaling it up with one hand while the other hand hits the power button at the top of the iPhone. This will kick you out of the application. Then re-launch the sample application and notice that the saucer is back at its original small scale.

The pinch gesture is the first multi-state gesture we have looked at in this chapter. The next gesture we are going to look at is the pan or drag gesture, which is another multi-state gesture. We will see that it behaves in very much the same way as the pinch gesture.

Pan (or Drag) Gesture

A pan gesture is the kind of gesture you use in the Maps application to view different parts of the map. A drag gesture is also used to unlock the phone or pull down the notification center. If you think about it, these gestures are really the same thing; the user puts their finger down, moves it in a mostly straight line, and then lifts it. The difference is in how the application responds. We call it a pan gesture when the view pans over some content (e.g., a map or image). We call it a drag when you move a UI component on the screen (e.g., the knob of a slider). Regardless of what we call it in our description, this gesture is implemented with the class UIPanGestureRecognizer (sorry, drag). Figure 8–4 shows an example using UIPanGestureRecognizer.

Figure 8–4. *Pan or drag example*

In Figure 8–4 we see three asteroids. Each asteroid can be dragged up or down along an invisible path. When dragging an asteroid, you can move your finger all the way to the right or the left. This makes it a lot easier to use, because the user does not have to be too carful about where she moves the finger once she gets the desired asteroid moving. This simple example starts with the header file for the class PanGestureController, as shown in Listing 8–17.

Listing 8–17. *PanGestureController.h*

```
@interface PanGestureController : GameController{
    NSMutableArray* asteroids;
    int asteroidIndex;
    CGPoint startCenter;
}
@property (nonatomic) float minYValue;
@property (nonatomic) float maxYValue;

-(void)panGesture:(UIPanGestureRecognizer*)panRecognizer;

@end
```

In Listing 8–17, we see the header for the class PanGestureController, which again
extends GameController. We keep an ordered list of the asteroids in the NSMutableArray
asteroids. We also have two variables keeping track of state for the duration of a
gesture: the int asteroidIndex and the CGPoint startCenter. The properties minYValue
and maxYValue specify how far each Asteroid may move up or down. The task
panGesture: is called by the UIPanGestureRecognizer when a pan gesture is detected.
The meat of this example is found in the implementation of the class
PanGestureController, let's start with the doSetup tasks, as shown in Listing 8–18.

Listing 8–18. *PanGestureController.m (doSetup)*

```
-(BOOL)doSetup{
    if ([super doSetup]){
        [self setGameAreaSize:CGSizeMake(320, 480)];
        [self setMinYValue:72.0f];
        [self setMaxYValue:480-[self minYValue]];

        asteroids = [NSMutableArray new];

        for (int i=0;i<3;i++){
            Asteroid* asteroid = [Asteroid asteroid:self];
            [[asteroid behaviors] removeAllObjects];

            [self addActor:asteroid];
            [asteroids addObject:asteroid];

            CGPoint center = CGPointMake(i*320.0/3.0+320.0/6.0, [self minYValue]);
            [asteroid setCenter:center];
        }

        UIPanGestureRecognizer* panRecognizer = [[UIPanGestureRecognizer alloc]
initWithTarget:self action:@selector(panGesture:)];
        [panRecognizer setMinimumNumberOfTouches:1];
        [actorsView addGestureRecognizer:panRecognizer];
        return YES;
    }
    return NO;
}
```

In Listing 8–18, we see the task doSetup for the class PanGestureController. The first
thing we do after specifying the size of the game area is specify the minYValue and the
maxYValue for the asteroids, so they can not move with 72 points of the top or bottom of

the screen. Next, we create our three Asteroid objects, add them to the scene, and add them to the NSMutableArray asteroids. Each Asteroid starts at the top of the screen.

The last thing we do in doSetup is create a UIPanGestureRecognizer, configure it to call panGesture:, and add it to actorsView. The next section explains how we respond to these gestures to create the desired behavior.

Responding to Pan Gestures

We have reviewed how to set up the class UIPanGestureRecognizer. Now we will explore how to respond to this gesture in the task panGesture:, as shown in Listing 8–19.

Listing 8–19. *PanGestureController.m (panGesture:)*

```
-(void)panGesture:(UIPanGestureRecognizer*)panRecognizer{

    if ([panRecognizer state] == UIGestureRecognizerStateBegan){
        CGSize gameAreaSize = [self gameAreaSize];
        CGPoint locationInView = [panRecognizer locationInView:actorsView];

        if (locationInView.x < gameAreaSize.width/3.0){
            asteroidIndex = 0;
        } else if (locationInView.x > gameAreaSize.width/3.0*2){
            asteroidIndex = 2;
        } else {
            asteroidIndex = 1;
        }

        Asteroid* asteroid = [asteroids objectAtIndex:asteroidIndex];
        startCenter = [asteroid center];
        [asteroid setAnimationPaused:YES];
    } else if ([panRecognizer state] == UIGestureRecognizerStateChanged){
        Asteroid* asteroid = [asteroids objectAtIndex:asteroidIndex];
        CGPoint locationInView = [panRecognizer locationInView:actorsView];

        CGPoint center = [asteroid center];
        center.y = locationInView.y;

        if (center.y < [self minYValue]){
            center.y = [self minYValue];
        }
        if (center.y > [self maxYValue]){
            center.y = [self maxYValue];
        }

        [asteroid setCenter:center];
    } else if ([panRecognizer state] == UIGestureRecognizerStateEnded){
        Asteroid* asteroid = [asteroids objectAtIndex:asteroidIndex];
        [asteroid setAnimationPaused:NO];
    } else if ([panRecognizer state] == UIGestureRecognizerStateCancelled){
        Asteroid* asteroid = [asteroids objectAtIndex:asteroidIndex];
        [asteroid setAnimationPaused:NO];
        [asteroid setCenter:startCenter];
    }

}
```

In Listing 8–19, we can see that the task panGesture: is broken up by the state of panRecognizer in the same way as with the pinch-out example. If the state of panRecognizer equals UIGestureRecognizerStateBegan, we calculate which of the three asteroids the gesture is closest to horizontally. Once we record which asteroid we are working with in asteroidIndex, we can get the active Asteroid object from the NSMutableArray asteroids and record its location in startCenter. Lastly, we stop it from spinning by setting animationPaused to NO.

In subsequent calls to panGesture the panRecognizer will have the state of UIGestureRecognizerStateChanged. When this happens, we simply set the Y value of the selected asteroid to the Y value of the gesture, using the task locationInView of panRecognizer.

Like our previous example, the gesture either terminates with either the state UIGestureRecognizerStateEnded or UIGestureRecognizerStateCancelled. In both cases, we want to restart the animation in the cancel state; we also want to revert the center of the selected asteroid back to startCenter.

I hope by this point a pattern is emerging as to how these gesture recognizers work. The basic pattern is to set up the recognizer to detect gestures on a UIView, configure the callback task to record the starting state of whatever is being manipulated (saucer, asteroid, and so on), apply changes to the scene, or rollback the changes in the case of cancelation. Let's look at the rotation gesture next.

Rotation Gesture

The rotation gesture is not as common as the previously covered gestures in the core iOS applications. In fact, I could not find a single example in the applications that ship with the iPhone, but I may be overlooking something. The rotation gesture, however, does show up in a number puzzle games, where you need to rotate an object this way or that. The rotation gesture is performed by putting two fingers on the screen and spinning your hand, like you are opening a bottle. Figure 8–5 shows a rotation example.

In Figure 8–5 we see a spaceship being rotated counter clockwise. We know the ship is being rotated counter clockwise because its starboard-maneuvering thruster is firing. If the ship were rotating clockwise, we would see the port maneuvering thruster firing. The class that is responsible for this example is RotationGestureController, whose header file is shown in Listing 8–20.

Gestures and Motion

Figure 8–5. *Rotating a spaceship with the rotation gesture*

Listing 8–20. *RotationGestureController.h*

```
@interface RotationGestureController : GameController{
    Viper* viper;
    float startRotation;
}
-(void)rotationGesture:(UIRotationGestureRecognizer*)rotationRecognizer;
@end
```

In Listing 8–20, we see the header file for the class RotationGestureController. We see that we have a reference to the spaceship, called viper. We also keep track of the starting rotation of the spaceship in the variable startRotation. The task rotationGesture: is called when a UIRotationGestureRecognizer recognizes a gesture. Let's take a quick look at the doSetup task for the class RotationGestureController so we understand how the UIRotationGestureRecognizer is set up. See Listing 8–21.

Listing 8–21. *RotationGestureController.m (doSetup)*

```
-(BOOL)doSetup{
    if ([super doSetup]){
        [self setGameAreaSize:CGSizeMake(320, 480)];

        viper = [Viper viper:self];
        [self addActor:viper];
```

```
        UIRotationGestureRecognizer* rotationRecognizer = [[UIRotationGestureRecognizer
alloc] initWithTarget:self action:@selector(rotationGesture:)];

        [actorsView addGestureRecognizer:rotationRecognizer];
        return YES;
    }
    return NO;
}
```

In Listing 8–21, we can see the doSetup task of RotationGestureController is very simple. We add a new Viper object and register a UIRotationGestureRecognizer. There should be no surprises at this point—we know that the task rotationGesture: will be called as the object rotationRecognizer detects touch events that can be interpreted as a rotation gesture. Let's take a look at the task rotationGesture:, as shown in Listing 8–22.

Listing 8–22. *RotationGestureController.m (rotationGesture:)*

```
-(void)rotationGesture:(UIRotationGestureRecognizer*)rotationRecognizer{

    if ([rotationRecognizer state] == UIGestureRecognizerStateBegan){
        startRotation = [viper rotation];
    } else if ([rotationRecognizer state] == UIGestureRecognizerStateChanged){

        float rotation = [rotationRecognizer rotation];
        float finalRotation = startRotation + rotation*2.0;

        if (finalRotation > [viper rotation]){
            [viper setState:VPR_STATE_CLOCKWISE];
        } else {
            [viper setState:VPR_STATE_COUNTER_CLOCKWISE];
        }

        [viper setRotation: finalRotation];

    } else if ([rotationRecognizer state] == UIGestureRecognizerStateEnded){
        [viper setState:VPR_STATE_STOPPED];
    } else if ([rotationRecognizer state] == UIGestureRecognizerStateCancelled){
        [viper setState:VPR_STATE_STOPPED];
        [viper setRotation:startRotation];
    }
}
```

In Listing 8–22, shows the rotationGesture: task for the class RotationGestureController. In this task, we perform the now familiar tasks based on the state of the rotationRecognizer. If the state is UIGestureRecognizerStateBegan, we simply record the starting rotation of actor viper. If the state is UIGestureRecognizerStateChanged, we get the rotation value from rotationRecognizer. The value of rotation is relative to where the user first placed his fingers. Since the value of rotation is relative to startRotation, we have to add rotation to startRotation to have the ship turn like a knob. We multiply the rotation by 2.0, to make the rotation faster and more responsive. This is only done because we might use this technique in an action game, where we want the user to be able to rotate the ship pretty quickly. Once the value of finalRotation is calculated, we see if it is bigger or smaller than the current rotation. This allows us to set the state of viper to either VPR_STATE_CLOCKWISE or VPR_STATE_COUNTER_CLOCKWISE.

When the gesture is over, we simply set the state of viper back to VPR_STATE_STOPPED. Additionally, if the gesture was cancelled, we reset the rotation of the ship back to the startRotation.

The next example is more similar to the tap example than to the previous example, because it handles what it is called a long press.

Long Press Gesture

The long press gesture is a gesture where the user touches a point on the phone and holds it for a bit. It is sort of like a tap, but longer—as the name implies. Users often invoke the long press gesture when they want to rearrange their apps—when you press an app to make them all start shaking it is a long press. You also use a long press when pressing a key on the keyboard that has sub-items. For example, open Safari and tap on the address field. Once the keyboard shows up, put your finger on the ".com" button. After a moment, five alternate top-level domains are displayed. For this example, we are going to combine taps with long presses to change which size bullet a ship fires, as shown in Figure 8–6.

Figure 8–6. *A ship shooting three different sizes of bullets*

In Figure 8–6, we see a ship toward the bottom of the screen. There are a number of circles above the ship; these are bullets that are shot out of the ship based on user gestures. The cluster of three circles toward the upper left are the basic bullets created from a simple tap. The circle closest to the bottom is a bigger bullet created when the user does a long press. If the user does a touch that is longer than two seconds, an even bigger bullet is fired, as can be seen in the upper right of Figure 8–6. Let's take a look at the header file for the class LongPressController in Listing 8–23.

Listing 8–23. *LongPressController.h*

```
@interface LongPressController : GameController{
    Viper* viper;
    NSDate* longStart;
}
-(void)tapGesture:(UITapGestureRecognizer*)tapRecognizer;
-(void)longPressGesture:(UILongPressGestureRecognizer*)longPressRecognizer;
-(void)fireBulletAt:(CGPoint)point WithDamage:(float)bulletSize;
@end
```

In Listing 8–23, we see the header for the class LongPressController, which shows that we have a reference to the ship as well as an NSDate that is used to keep track of when a long press gesture starts. The task tapGesture: is used when the user simply taps the screen to fire a small bullet. The task longPressGesture: is called when the user performs a long press gesture to fire a larger bullet. The last task is used to create a Bullet actor and add it to the scene. Let's start by looking at the doSetup task of LongPressController. See in Listing 8–24.

Listing 8–24. *LongPressController.m (doSetup)*

```
-(BOOL)doSetup{
    if ([super doSetup]){
        [self setGameAreaSize:CGSizeMake(320, 480)];

        viper = [Viper viper:self];
        [self addActor:viper];

        CGPoint center = [viper center];
        center.y = [self gameAreaSize].height/5.0*4.0;
        [viper setCenter:center];

        UITapGestureRecognizer* tapRecognizer = [[UITapGestureRecognizer alloc]
initWithTarget:self action:@selector(tapGesture:)];

        [actorsView addGestureRecognizer:tapRecognizer];

        UILongPressGestureRecognizer* longPressRecognizer =
[[UILongPressGestureRecognizer alloc] initWithTarget:self
action:@selector(longPressGesture:)];
        [longPressRecognizer setMinimumPressDuration:1.0f];

        [actorsView addGestureRecognizer: longPressRecognizer];

        return YES;
    }
    return NO;
}
```

In Listing 8–24, we see task doSetup. After we perform the usual steps of setting the game area size and adding the actor viper, we register two gesture responders. The first gesture recognizer is a UITapGestureRecognizer that will call tapGesture:. The second gesture recognizer is a UILongPressGestureRecognizer that is configure to call longPressGesture:. The UILongPressGestureRecognizer has the property minimumPressDuration set to 1.0f. This means that the user must hold the press for one second before this UILongPressGestureRecognizer will consider the touches a long press. Both gesture recognizers are added to actorsView.

Responding to the User

In the following section, we will look at how bullets are added based on which gesture was triggered. Let's get the tap gesture out of the way first. The implementation of tapGesture: is shown in Listing 8–25.

Listing 8–25. *LongPressController.m (tapGesture:)*

```
-(void)tapGesture:(UITapGestureRecognizer*)tapRecognizer{
    [self fireBulletAt:[tapRecognizer locationInView:actorsView] WithDamage:10];
}
```

In Listing 8–25, we see that every time a tap gesture is recognized, we simply call fireBulletAt:WithDamage: specifying the location of the tap and the damage value of 10. The more interesting interaction is found is the implementation of longPressGesture:, as shown in Listing 8–26.

Listing 8–26. *LongPressController.m (longPressGesture:)*

```
-(void)longPressGesture:(UILongPressGestureRecognizer*)longPressRecognizer{

    if ([longPressRecognizer state] == UIGestureRecognizerStateBegan){
        [viper setState:VPR_STATE_TRAVELING];
        longStart = [NSDate date];

    } else if ([longPressRecognizer state] == UIGestureRecognizerStateEnded){

        NSDate* now = [NSDate date];

        float damage = 20;
        if ([now timeIntervalSinceDate:longStart] > 1.0f){
            damage = 30;
        }

        [self fireBulletAt:[longPressRecognizer locationInView:actorsView]
WithDamage:damage];
        [viper setState:VPR_STATE_STOPPED];
    }
}
```

In Listing 8–26, we use the state of longPressRecognizer to decide what should be done. If the gesture has been started, we record the time in the variable longStart; we also set the state of the viper object to VPR_STATE_TRAVELING, which causes the thrust of the ship to be shown. If you have played with the demo application, you will notice that the thrust comes on after the one-second delay associated with the gesture. This means that `longPressGesture:` does not get called until this minimum amount of time has passed.

If the state of `longPressRecognizer` is `UIGestureRecognizerStateEnded`, then we know the user has lifted her finger. When this happens, we see if more than a second has passed since the gesture was started. If so, we upgrade the damage from 20 to 30. This means that if the user holds her finger down for a second, the next bullet will do 20 damage, and if she holds it down for two seconds, the damage will be 30. When the long press gesture is over, we set the state of the viper back to VPR_STATE_STOPPED and fire the bullet with the appropriate damage.

Adding the Bullet

Let's take a look at the implementation of `fireBulletAt:WithDamage:` to complete this example. See Listing 8–27.

Listing 8–27. *LongPressController.m (fireBulletAt:WithDamage:)*

```
-(void)fireBulletAt:(CGPoint)point WithDamage:(float)damage{
    Bullet* bullet = [Bullet bulletAt:[viper center] TowardPoint:point];
    [bullet setDamage:damage];
    [self addActor:bullet];
}
```

In Listing 8–27, we see that we simply create a new `Bullet` actor that is configured to travel toward a specified point and add it the scene.

Swipe Gesture

The swipe gesture is probably most recognizable as the gesture used to switch the set of apps you see on the home screen. This gesture is often used to switch context, like with the home screen. Other examples include the four-finger swipe used to switch the foreground application on the iPad 2. It is sort of like pawing your way through your running applications. In our example, we are going to use swipe gestures to add comets to the scene, as shown in Figure 8–7.

In Figure 8–7, we see several comets on the screen. Each comet is created by a swipe gesture that determines where the comet starts and in which direction it travels. For example, a swipe to the right causes a comet to be created on the left and travel to the right. Let's take a look at the doSetup task of the SwipeGestureController class in Listing 8–28.

Figure 8–7. *Swipes create comets*

Listing 8–28. *SwipeGestureController.m (doSetup)*

```
-(BOOL)doSetup{
    if ([super doSetup]){
        [self setGameAreaSize:CGSizeMake(320, 480)];

        UISwipeGestureRecognizer* down = [[UISwipeGestureRecognizer alloc]
initWithTarget:self action:@selector(swipeGesture:)];
        [down setDirection:UISwipeGestureRecognizerDirectionDown];
        [actorsView addGestureRecognizer:down];

        UISwipeGestureRecognizer* up = [[UISwipeGestureRecognizer alloc]
initWithTarget:self action:@selector(swipeGesture:)];
        [up setDirection:UISwipeGestureRecognizerDirectionUp];
        [actorsView addGestureRecognizer:up];

        UISwipeGestureRecognizer* left = [[UISwipeGestureRecognizer alloc]
initWithTarget:self action:@selector(swipeGesture:)];
        [left setDirection:UISwipeGestureRecognizerDirectionLeft];
        [actorsView addGestureRecognizer:left];

        UISwipeGestureRecognizer* right = [[UISwipeGestureRecognizer alloc]
initWithTarget:self action:@selector(swipeGesture:)];
        [right setDirection:UISwipeGestureRecognizerDirectionRight];
        [actorsView addGestureRecognizer:right];
```

```
        return YES;
    }
    return NO;
}
```

In Listing 8–28, we see the code that registers the gesture recognizers for this example. We see that we create a UISwipeGestureRecognizer for each direction we want to listen for. The only surprise in this example is that it requires four different gesture recognizers to achieve what we want. UISwipeGestureRecognizer does allow you to specify a single recognizer for multiple directions. However, when responding to a swipe gesture, there is no way to tell which direction the gesture was in. Only by creating a recognizer for each direction can we tell which direction the user's gesture was in.

Let's look at the task swipeGesture:. See Listing 8–29.

Listing 8–29. *SwipeGestureController.m (swipeGesture:)*

```
-(void)swipeGesture:(UISwipeGestureRecognizer*)swipeRecognizer{
    CGSize gameSize = [self gameAreaSize];

    UISwipeGestureRecognizerDirection direction = [swipeRecognizer direction];
    CGPoint locationInView = [swipeRecognizer locationInView:actorsView];


    CGPoint center = CGPointMake(0, 0);
    float directionInRadians = DIRECTION_DOWN;

    if (direction == UISwipeGestureRecognizerDirectionRight){
        center.x = -20;
        center.y = locationInView.y;
        directionInRadians = DIRECTION_RIGHT;
    } else if (direction == UISwipeGestureRecognizerDirectionDown){
        center.x = locationInView.x;
        center.y = -20;
        directionInRadians = DIRECTION_DOWN;
    } else if (direction == UISwipeGestureRecognizerDirectionLeft){
        center.x = gameSize.width+20;
        center.y = locationInView.y;
        directionInRadians = DIRECTION_LEFT;
    } else if (direction == UISwipeGestureRecognizerDirectionUp){
        center.x = locationInView.x;
        center.y = gameSize.height+20;
        directionInRadians = DIRECTION_UP;
    }

    Comet* comet = [Comet comet:self withDirection:directionInRadians andCenter:center];
    [self addActor:comet];
}
```

In Listing 8–29, we see the code that is called when a swipe gesture is recognized. To create the effect we want, we are simply going to create a new Comet actor with the correct location and direction. To figure out where the Comet should be placed and in what direction it travels, we simply look at the direction property of swipeRecognizer.

We are not going to go into the details of how the Comet class is implemented; we have discussed that at length already. If you are really interested, please review the source code provided.

We have looked at the built-in gesture recognizers available in iOS and have explored the basics of how they work. In the next section, we will explore how movements of the device can be incorporated into an application.

Interpreting Device Movements

Starting with the iPhone 4, iOS devices have an accelerometer and a gyroscopic sensor built into them. There are several ways to use these sensors in an application. The two most common ways are to simply respond to a motion event, which corresponds to shaking the device. The other and more fine-tuned way is to access the orientation data of these sensors directly through the UIAccelerometer class. Let's take a look at the shake example first.

Responding a to Motion Event (Shaking)

Whenever an iOS device is shaken, a motion event is generated for the application to consume. Personally, I find using the shake in an application to be super annoying; however, some people think it is a good idea, so let's look at how we can take advantage of this feature. Figure 8–8 shows our example.

Figure 8–8. *Asteroids breaking apart from a device shake*

In Figure 8–8, we see a number of asteroids. These asteroids used to be a single asteroid, but a few shake events caused it to break apart. The class for this example is called ShakeController and extends GameController. There is nothing particularly interesting about its header file, so let's just jump to its doSetup task, shown in Listing 8–30.

Listing 8–30. *ShakeController.m (doSetup)*

```
-(BOOL)doSetup{
    if ([super doSetup]){
        [self setGameAreaSize:CGSizeMake(320, 480)];

        NSMutableArray* classes = [NSMutableArray new];
        [classes addObject:[Asteroid class]];

        [self setSortedActorClasses:classes];

        return YES;
    }
    return NO;
}
```

In Listing 8–30, we see the task doSetup that configures the ShakeController, marking Asteroid as a class of Actor we will want to get access to later. We add our Asteroid in the task updateScene. We do this since we want to only add a new Asteroid when there are no Asteroid objects in the scene, as can be seen in Listing 8–31.

Listing 8–31. *ShakeController (updateScene)*

```
-(void)updateScene{
    if ([[self actorsOfType:[Asteroid class]] count] ==0 ){
        Asteroid* asteroid = [Asteroid asteroid:self];
        [self addActor:asteroid];
    }
    [super updateScene];
}
```

In Listing 8–31, we see the updateScene task that is called in order to move the game forward through each step of the animation. In this task, we check to see if there are Asteroid objects in the scene; if there are none, we go ahead and add one. Lastly, we call the super implementation of updateScene to drive the game forward.

Now the question is, how do we respond to shake events so we can blow up the asteroid? Informing the application that the top view can become the "first responder" and implement the appropriate method to respond to a motion event achieves this. A UIView that is the first responder is a UIView that gets the first crack at interpreting user input. There are two steps to getting a view to be first responder. First you have to inform the application that the view property of a UIViewController can be the first responder, and then you have to make a request to become first responder. Listing 8–32 shows the code to accomplish this.

Listing 8–32. *ShakeController (first responder related tasks)*

```
-(BOOL)canBecomeFirstResponder {
    return YES;
}

-(void)viewDidAppear:(BOOL)animated {
    [super viewDidAppear:animated];
    [self becomeFirstResponder];
}

- (void)viewWillDisappear:(BOOL)animated {
    [self resignFirstResponder];
    [super viewWillDisappear:animated];
}
```

In Listing 8–32, we see three tasks. The first task, canBecomeFirstResponder, simply informs the application that the view property of this UIViewController is eligible to be first responder, and that this UIViewController will handle its events. The next two tasks are part of a UIViewControllers life cycle: viewDidAppear: is called when this UIViewController instance's view property is displayed, when this happens we can become first responder by calling becomeFirstResponder. Conversely, when this UIViewsController is removed from the screen, we no longer want this UIViewController to be first responder, so we call resignFirstResponder. These three tasks simply make this class eligible to receive motion events, these are handled in the implementation of the task motionBegan:withEvent: and motionEnded:withEvent:. We only need to implement one of these in our example, the implementation of motionBegan:withEvent: is shown in Listing 8–33.

Listing 8–33. *ShakeController.m (motionBegan:withEvent:)*

```
- (void)motionBegan:(UIEventSubtype)motion withEvent:(UIEvent *)event
{
    if (motion == UIEventSubtypeMotionShake)
    {
        for (Asteroid* asteroid in [self actorsOfType:[Asteroid class]]){
            [asteroid doHit:self];
        }
    }
}
```

In Listing 8–33, we see the task motionBegan:withEvent:, where we first check that the motion subtype is a UIEventSubtypeMotionShake. If it is, we simply iterate over all of the asteroids in the scene and call doHit: on them. The task doHit: is discussed in Chapter 7.

As we can see responding to shake events is pretty simple, very much like responding to touch events. I find it a little weird that shake events are not implemented as a gesture. I suspect the reason is that shake events have been around before the many different gesture recognizers made it into the SDK.

Responding to Accelerometer Data

As a developer, being able to know the orientation of the device is really cool to me. It allows things like augmented reality to be possible, as well as a bunch of other cool things. The last example in this chapter will use the accelerometer to manipulate actors in our scene. I can't promise it will be as cool as an augmented reality game, but it should help to pave the way for other developers. Figure 8–9 shows the accelerometer example being run.

Figure 8–9. *Multiple screen of accelerometer demo*

In Figure 8–9, we see several screenshots of this example. In this example, the location of the viper is based on the x and y accelerometer values, so when the iPhone is flat on its back, the spaceship should be in the center of the game area. When the device is tilted to the left, the ship moves to the left. When the device is tilted up, the ship will move up, and so on. Figure 8–8 highlights how jumpy the accelerometer data is. These were all taken while I was trying to hold the device flat (though I was trying to take screenshots at the same time). For a real application, the accelerometer data would need to be massaged.

The class responsible for our example is the class `AccelermoterController`, the header of which is shown in Listing 8–34.

Listing 8–34. *AccelerometerController.h*

```
@interface AccelerometerController : GameController<UIAccelerometerDelegate>{
    Viper* viper;
}
@end
```

In Listing 8–34, we see the header for the class AccelerometerController. The things to note are that the class AccelerometerController conforms to the protocol UIAccelerometerDelegate, and that it has a reference to the Viper that will be moving around the screen. Conforming the protocol allows AccelerometerController to receive data from the accelerometer in real time, as we will see in a bit. Let's first look at how setup the class AccelerometerController so it receives data, as shown in Listing 8–35.

Listing 8–35. *AccelerometerController.m (doSetup)*

```
-(BOOL)doSetup{
    if ([super doSetup]){
        [self setGameAreaSize:CGSizeMake(320, 480)];

        viper = [Viper viper:self];
        [self addActor:viper];

        UIAccelerometer*  theAccelerometer = [UIAccelerometer sharedAccelerometer];

        theAccelerometer.updateInterval = 1 / 50.0;
        theAccelerometer.delegate = self;

        return YES;
    }
    return NO;
}
```

In Listing 8–35, we do the usual setup, specifying the game size area and adding the viper that we will be moving around with accelerometer data. To start receiving accelerometer data, we have to get access to the singleton UIAccelerometer associated with the running application. To do this, we use the static task sharedAccelermoter of the class UIAccelermoter, and then we simply set self as the delegate. Another thing to note is that we have to specify the frequency at which we get data. In this case, we are requesting data 50 times a second, which is plenty fast for real-time input. The data is actually received by a call to accelerometer:didAccelerate:, as shown in Listing 8–36.

Listing 8–36. *AccelerometerController.m (accelerometer:didAccelerate:)*

```
- (void)accelerometer:(UIAccelerometer *)accelerometer didAccelerate:(UIAcceleration
*)acceleration{
    CGSize size = [self gameAreaSize];

    UIAccelerationValue x, y, z;
    x = acceleration.x;
    y = acceleration.y;
    z = acceleration.z;
    NSLog(@"x = %f y = %f z = %f", x, y, z);

    CGPoint center = CGPointMake(size.width/2.0 * x + size.width/2.0, size.height/2.0 *
y + size.height/2.0);
    [viper setCenter:center];
}
```

In Listing 8–36, we simply get the x, y, and z values from the UIAcceleration acceleration. These values range from -1.0 to 1.0, based on the orientation of the device. Because the x value corresponds to the device being tilted left or right (in portrait), we can simply use the value of x to specify the x value of our new center value. We do the same the y value, and set the center property of viper.

This is a very simple example, but hopefully it will get you headed down the right track. It seems to me that getting basic functionality working is pretty simple, as in this example, but getting more complex user interactions just right is considerably more work. I recommend reading the document entitled "Event Handling Guide for iOS," which is available at http://developer.apple.com.

Summary

In this chapter, we explored how to respond to basic touch events. We also looked at how the subclasses of UIGestureResponder can be used to interpret those touch events as concise gestures such as pinches, long presses, rotations, and more. In the last section, we looked at two ways to use the accelerometer to control game actors. The first method allowed us to simply respond to a shake event. The second technique showed how we could get direct access to accelerometer data to control our application, laying the groundwork for motion-based gestures more complex than a simple shake.

Chapter 9

Game Center and Social Media

One of the key ways to market your game these days is to integrate your application with one or more social media services. The idea is that users will promote your game as they play and share their experience with their friends. To get users to play your game more often, it is also very common to provide a mechanism for users to compare their scores with other players. Providing a way for users to track their scores against others and to post their progress is so common in game development that Apple provides libraries to support these requirements.

The primary service provided by Apple is called Game Center, which is implemented in the library called GameKit. Game Center is a service that allows users to compare their scores with other players on what is called a leaderboard. Game Center also allows users to reach what are called achievements. Achievements are defined by each game and are intended as incentives to keep people playing. For example, if you have an achievement that is hard to achieve, people will want to earn it to set them apart from their friends playing the same game. Game Center also provides a mechanism for multiplayer games, but we will focus more on the social aspects of its services in this chapter.

To broadcast a player's progress in a game, it is common to use an existing social media site such as Twitter or Facebook, so the user's friends will be notified. A new feature in iOS 5 is the inclusion of a built-in Twitter account. In this chapter, we will look at how to use the classes that expose a Twitter account to send a tweet on behalf of the user. Although Facebook is not built into iOS 5, there is an excellent library provided by Facebook to facilitate an experience almost as integrated as that of Twitter. We will look at how to use Facebook on iOS to facilitate user authenticating as well as posting on his wall.

The examples in this chapter are taken from the project Belt Commander. We will continue using what you learn over the following pages as examples for the final two chapters, as we explore a complete game.

Game Center

Game Center, as the user sees it, is an application on an iOS device where information about a user's games is aggregated. This information includes a place to log in and manage your account, a list of friends, and a list of games. For each listed game, there are a number of leaderboards for displaying scores, as well as a list of the achievements available in the game. Figure 9–1 shows a typical Game Center screen.

Figure 9–1. *Game Center showing games*

In Figure 9–1, we see nine installed games that support Game Center. Select a game to view its leaderboards and any achievements the user has earned.

From a developer's perspective, Game Center is a set of classes used to interact with Game Center to define the leaderboards and achievements. Collectively, these classes are called GameKit, which is available as part of your standard iOS SDK installation. Once GameKit is integrated with your game, will be able to sign the user in and perform GameKit operations on her behalf. Figure 9–2 shows the user being signed into GameKit.

Figure 9–2. *Signing into GameKit*

In Figure 9–2, we see the Welcome screen for the game Belt Commander. At the top, we see a notification from Game Center letting the user know that this game is using GameKit with the user account displayed. We also see the word "sandbox." This is informing us that we're not using the live version of Game Center, but rather the development version. For the most part, you don't have to worry about this; while developing, you will be using the sandbox exclusively.

To enable GameKit in your project, you obviously use classes from GameKit, but you also have to tell Apple that you want GameKit support for a given game. The following section describes how to set up GameKit in iTunes Connect.

Enabling Game Center in iTunes Connect

Game Center is not just an app or set of libraries on your iOS device. It is also a set of web services hosted by Apple that provide much of the functionality. In order for your game to use these services, you have to configure your application. You will have to create an App ID in the provisioning portal, configure Xcode to use the resultant Bundle Identifier, and configure your application in iTunes Connect to use Game Center. Let's assume we are making a new game and start with the Provisioning Portal, as shown in Figure 9–3.

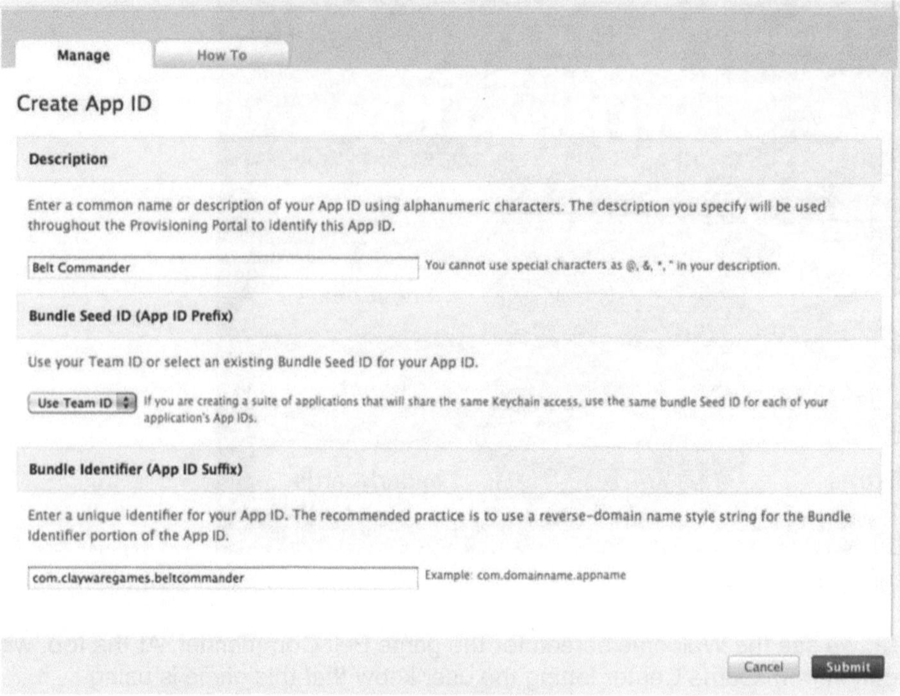

Figure 9–3. *Creating an App ID in the Provisioning Portal*

In Figure 9–3, we see the form used to create an App ID. The top field labeled Description can be anything you want to remind you why you created this App ID. You can select any Bundle Seed ID you want, but I have found an important reason not to just use the default: selecting the default Bundle Seed ID removes a number configuration steps when working with Xcode.

The last item is the Bundle Identifier. This is used to identify your application at runtime. For games that are meant to support Game Center, you have to specify a Bundle Identifier without a wildcard. Once you press Submit, you should open your game project in Xcode.

The Bundle Identifier must be set on Xcode, so iOS knows how to associate the binary running on your device to the application defined by iTunes connect. Figure 9–4 shows where to set the Bundle Identifier.

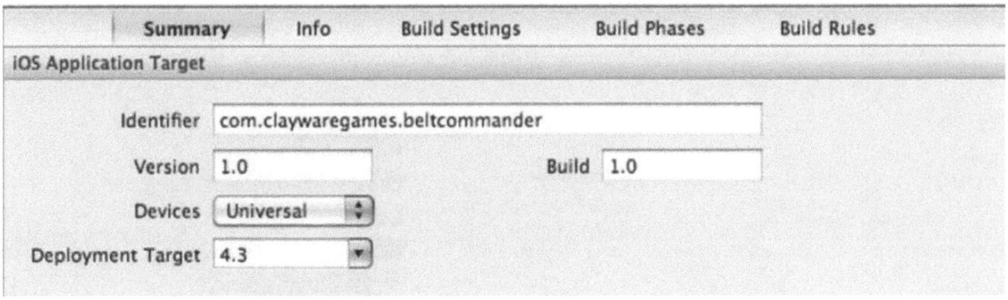

Figure 9–4. *Bundle Identifier in Xcode*

In Figure 9–4, the Identifier field contains our Bundle Identifier in Xcode. This value, in combination with the version number, identifies to iTunes Connect which application is running. Once you have set the Bundle Identifier, you have to create a new application in iTunes connect, as shown in Figure 9–5.

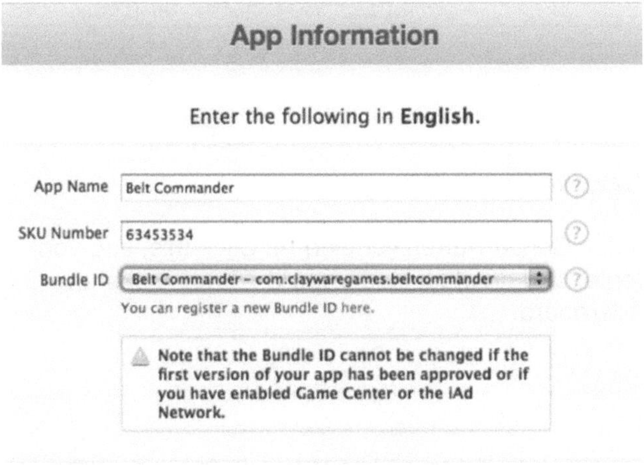

Figure 9–5. *New application in iTunes Connect*

In Figure 9–5, we see the first step in creating a new application in iTunes Connect. We have specified the App name and selected an SKU number. The SKU number can be anything you want; it is included so the producer of the game can track sales in their internal systems. We also see that we can select the Bundle ID from a pull-down menu; this is the value we specified when we created a new App ID. There are a few more steps required to create a new application in iTunes Connect, but they are just filling in information about the app and are not required steps to getting Game Center working. You will be asked to provide a description, copyright information, and other details. All of the information added at the lest step can be skipped and added later.

Once the application is created, inspect its details and find the Manage Game Center button, as shown in Figure 9–6.

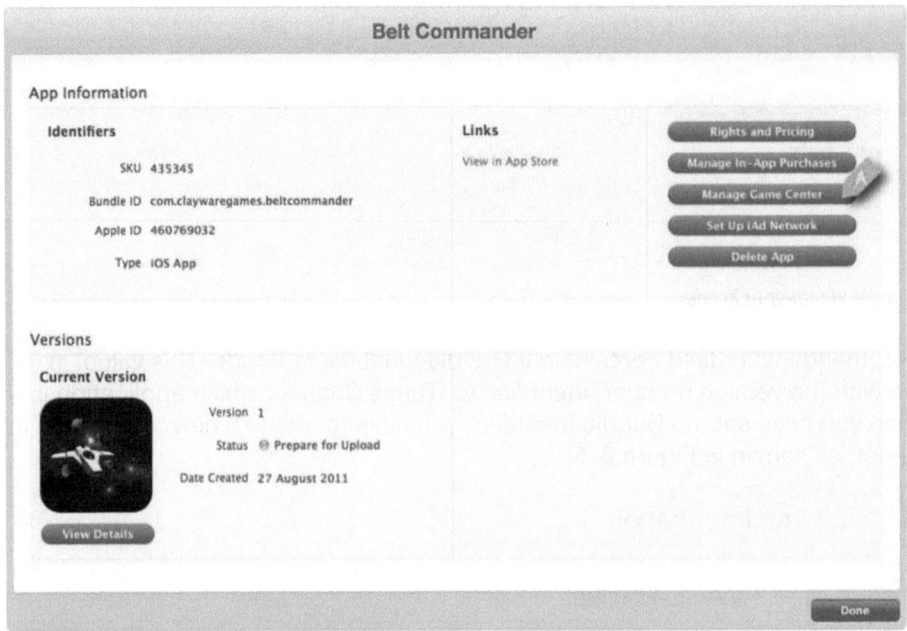

Figure 9–6. *Configuration buttons for an application in iTunes Connect*

In Figure 9–6, we see the Manage Game Center button (labeled A). Click this, and you will be prompted to enable Game Center and create a leaderboard. Figure 9–7 shows the page for creating or editing a leaderboard.

Figure 9–7. *Creating or editing a leaderboard*

In Figure 9–7, we see the web page in iTunes Connect for editing and creating a leaderboard. At item A, we see a name for the leaderboard. This name is not the display name—it is for internal use only and can be anything you want. Item B indicates the ID for the leaderboard. This is the value used in code when referring to the leaderboard. Items C and D indicate how the scores in this leaderboard should be organized. We have chosen to display our scores as an integer value (with no decimal places), and we consider higher scores better. Item E shows that we have added display information for one language. New languages can be added by clicking the Add Language button and specifying the display name of the leaderboard, as well as indicating that the values are called "points" in our case. Creating or editing an achievement is very similar to working with a leaderboard, as shown in Figure 9–8.

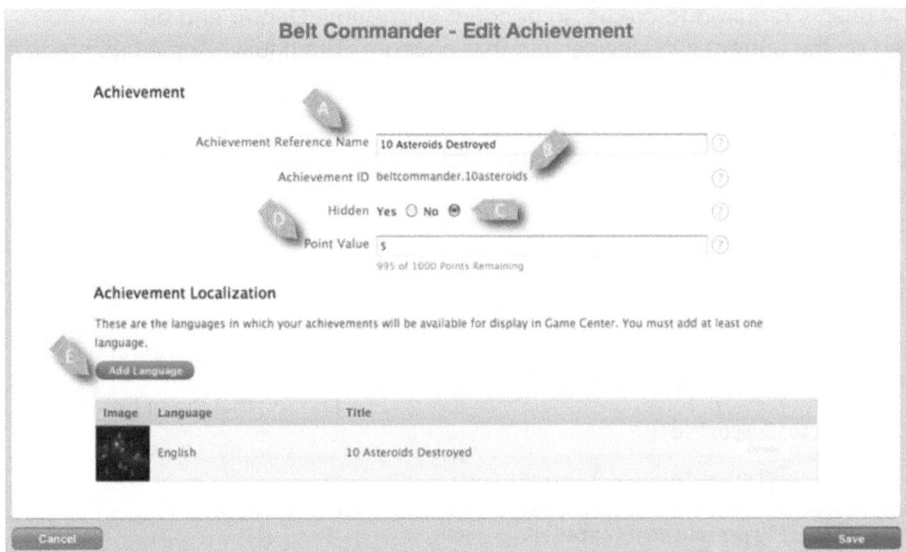

Figure 9–8. *Creating or editing an achievement*

In Figure 9–8, we see the web page for creating and editing an achievement in iTunes Connect. Item A shows the internal name used for the achievement. Item B is where the ID for the achievement is displayed. Item C indicates that this achievement is not hidden. By default, achievements are viewable to users even if they have not achieved them. However, it can be fun to specify some achievements that are listed, so the user is surprised when they achieve them. Item E shows that we have specified the display text for the English language. Of course, text for other languages can be added by clicking the Add Language button.

We have looked at how to create a new App ID and a new application in iTunes Connect. We also briefly looked at the screens required to create a leaderboard and an achievement. Now, let's look at the code that takes advantage of these new resources.

Using Game Center in Your Game

It is surprisingly easy to work with GameKit to start using the leaderboard and achievement features once Game Center is set up to include them. The following sections will discuss how to connect to the user's Game Center account, submit his scores to a leaderboard, and earn an achievement.

Enabling Game Center for a User

When a game is Game Center-enabled, the user must still opt into the Game Center features. The user is required to create an account with Game Center and be authenticated on the current iOS device with that account. Thankfully, Apple has made it simple to get going with Game Center.

To get started with Game Center, all apps must get access to an instance of GKLocalPlayer that represents the local player and authenticate them. Listing 9–1 shows the canonical example of doing this.

Listing 9–1. *RootViewController.m (initGameCenter)*

```
-(void)initGameCenter{
    Class gkClass = NSClassFromString(@"GKLocalPlayer");

    BOOL iosSupported = [[[UIDevice currentDevice] systemVersion] compare:@"4.1"
options:NSNumericSearch] != NSOrderedAscending;

    if (gkClass && iosSupported){

        localPlayer = [GKLocalPlayer localPlayer];
        [localPlayer authenticateWithCompletionHandler:^(NSError *error) {
            if (localPlayer.authenticated){
                [leaderBoardButton setEnabled:YES];
            } else {
                [leaderBoardButton setEnabled:NO];
            }
        }];
    }
}
```

In Listing 9–1, we see the task initGameCenter of the class RootViewController. This task is called once at the beginning of the application launch. It first checks to see if GameKit is available on this iOS device by looking for the class GKLocalPlayer. The tasks also checks to see if the version of iOS is recent enough to support Game Center. If yes, we get a pointer to the local player by calling localPlayer on the class GKLocalPlayer.

To authenticate our local player, we simply call authenticateWithCompletionHandler: and pass a block that will be executed asynchronously when the user has authenticated (or cancelled). A block is simply a way of declaring a function that can be passed around as value. In this case, we are using a block in the same way we would use an anonymous class on a language like Java. The first time this code is run, the user will see something like Figure 9–9.

Figure 9–9. *First-time authentication with Game Center*

In Figure 9–9, we see the result of calling GameKit's authentication task. The user is prompted to either sign in to Game Center with an existing account, create a new account, or cancel. If the user cancels enough times, the application will no longer display this dialog. The exact number of times that the user must cancel before this happens seems to change with different releases. With iOS 5, it is approximately three times. Once the user has successfully authenticated, the application is now connected to Game Center for that user account. When the application is run again, calls to authenticateWithCompletionHandler: will call the handler code without pestering the user to sign in again. The user will see something like Figure 9–2.

Once Game Center is enabled, we will want to start taking advantage of its features. The following section looks at how to report a score to a leaderboard.

Submitting Scores to a Leaderboard

One of the best features that Game Center provides is a built-in mechanism for tracking a user's high scores across devices. It allows users to compare their scores with their friends' scores, as well as with other players globally. The code for submitting a score is straightforward, as shown in Listing 9–2.

Listing 9–2. *RootViewController.m (notifyGameCenter)*

```
-(void)notifyGameCenter{
    if ([localPlayer isAuthenticated]){
        GKScore* score = [[GKScore alloc] initWithCategory:@"beltcommander.highscores"];
        score.value = ([beltCommanderController score];

        [score reportScoreWithCompletionHandler:^(NSError *error){
            if (error){
                //handle error
            }
        }];
    }
}
```

In Listing 9–2, we see the task notifyGameCenter, which is called every time a game ends. In this task, we check to see if the localPlayer is authenticated. If so, we simply create a GKScore object, specifying the ID of the leaderboard to which we are submitting the score, in this case "beltcommander.highscores". Once we have a reference to the GKScore object, we set the score property to the score from the last game. The property score can by any type of number—in this case it is a long. Lastly, we call reportScoreWithCompletionHandler: and pass a block to handle any error. In our example, we don't do anything with the error.

> **NOTE:** If the user is offline or there was some sort of network issue, GameKit will automatically resubmit scores when it is run in the future, ensuring that a player's hard work is properly recorded.

Figure 9–10 shows a score on the leaderboard.

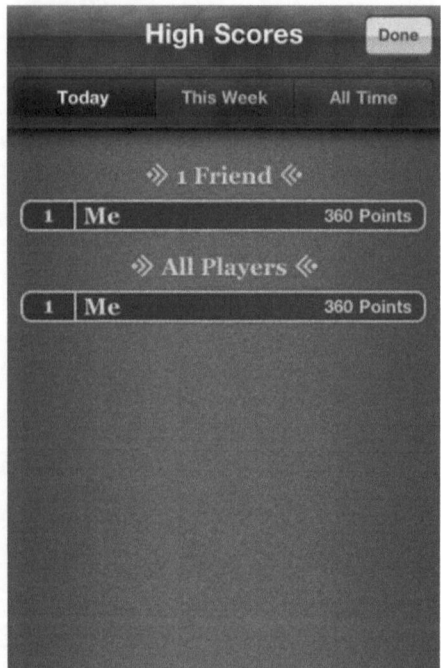

Figure 9–10. *A leaderboard with a sample score*

In Figure 9–10, we see a score displayed on a Leaderboard. There is only one score at the moment, because the screenshot was taken during development of the game. At the top, we can see three options for viewing high scores: Today, This Week, or All Time. The component that is displaying the high scores is a pre-built component called GKLeaderboardViewController that comes with GameKit; the code to bring up this component is shown in Listing 9–3.

Listing 9–3. *RootViewController.m (leaderBoardClicked:)*

```
- (IBAction)leaderBoardClicked:(id)sender {
        GKLeaderboardViewController* leaderBoardController =
[[GKLeaderboardViewController alloc] init];
        leaderBoardController.category = @"beltcommander.highscores";
        leaderBoardController.leaderboardDelegate = self;
        [self presentModalViewController:leaderBoardController animated:YES];
}
```

In Listing 9–3, we see the task leaderBoardClicked:, which is called when the user clicks the leaderboard button. This class is called on an instance of RootViewController, which is the main UIViewController for our example game Belt Commander. We don't have to know much about RootViewController except that it is a UIViewContoller and conforms to GKLeaderboardViewControllerDelegate. To display the view shown in Figure 9–10, first create a GKLeaderboardView and set the category property to the ID of our leaderboard. Then call presentModalViewController:animated: on self, which will display the Leaderboard view. In order to remove the leaderboard when the user clicks Done, set self as the leaderbordDelegate, which means the task leaderboardViewControllerDidFinish: from GKLeaderboardViewControllerDelegate will be called when the user is done. The implementation leaderboadViewControllerDidFinish: should be as follows in Listing 9–4.

Listing 9–4. *RootViewController.m (leaderboardViewControllerDidFinish:)*

```
- (void)leaderboardViewControllerDidFinish:(GKLeaderboardViewController
*)viewController{
    [self dismissModalViewControllerAnimated:YES];
}
```

In Listing 9–4, we see our implementation of leaderboardViewControllerDidFinish:, where we simply call dismissModalViewControllerAnimated: to remove the leaderboard view from the screen.

As you can see, working with scores and leaderboards with GameKit is pretty simple—it takes less than 10 lines for code to get things going. Next, let's look at working with achievements and seeing how they are awarded to a user.

Awarding Achievements

Achievements are little milestones that are reached throughout a game. They are intended to encourage the user to continue playing by indicating that they have achieved something. In addition to providing a sense of progress, you can also make the player feel clever by awarding secrete achievements. Previously, we looked at how to add an achievement in iTunes Connect (see Figure 9–8). In our sample game, Belt Commander, the player destroys wave after wave of asteroids. We want to give the user an achievement when he destroys his first 10 asteroids—an easy achievement, but it will remind the user early in the game that there are other achievements to earn. Figure 9–11 shows the achievement in the Game Center app.

Figure 9–11. *First achievement in the Game Center app*

In Figure 9–11, we see a screenshot from the Game Center app displaying the achievements for Belt Commander. We can see that the achievement in this view was awarded. Listing 9–5 shows the code that awards the achievement.

Listing 9–5. *BeltCommanderController.m (checkAchievements)*

```
-(void)checkAchievements{
    if (asteroids_destroyed >= 10){

        GKAchievement* achievement = [[GKAchievement alloc]
initWithIdentifier:@"beltcommander.10asteroids"];
        achievement.percentComplete = 100.0;

        [achievement reportAchievementWithCompletionHandler:^(NSError *error) {
            if (error){
                // report error
            }
        }];
    }
}
```

In Listing 9–5, we see the task checkAchievements from the class BeltCommanderController, which is responsible for handling the in-play game logic. The first thing we do is check if the asteroids_destroyed is bigger or equal to 10. The variable asteroids_destroyed is simply incremented whenever an asteroid in the game is destroyed. If asteroids_destroyed is at least 10, we get a reference to a GKAchievement by passing the achievement's ID (as defined in iTunes Connect) to the initWithIdentifier task of GKAchievement. Once we have a reference, we simply assign 100 to the property percentComplete, call reportAchievementWithCompletionHander, and we are done.

There are a few things to note about achievements. First, as shown in Listing 9–5, achievements can be partially complete. We could have set percentComplete to 30 or 50. This is an important feature for long-running achievements. Say, for example, our achievement was to destroy 10,000 asteroids. We would want to let the user know how she has progressed to date or she may never pursue it.

Another thing to note about achievements is that each has a value. Our achievement is worth 5 points, as shown in Figure 9–11. The value of an achievement can be whatever the developer wants. However, you are limited to a total of 1,000 points of achievements for a given game. This is so a user knows that he has achieved all the achievements for a given game when he hits 1,000 achievement points. It prevents the user from having to relearn the relative values of an achievement for a given game. A user knows that a 1- or 5-point achievement is not really a big deal, but that a 100-point achievement is a real accomplishment.

We have looked at Game Center and GameKit and how to work with leaderboards and achievements. As mentioned, GameKit can also be used to coordinate multiplayer games. We are not going to get into that topic here, as I believe it is too advanced for this book. However, I encourage you to follow up on GameKit's support for multiplayer games, as it should simplify your life considerably. Next, let's look at the Twitter integration new to iOS 5.

Twitter Integration

Love it or hate it, Twitter exists. As a mobile application developer, it is becoming more and more common for clients to ask for social media integration in their apps. On the short list of everyone's social services is Twitter. Twitter provides a very nice and useful REST service that can be used to tweet on behalf of a user. New to iOS 5 is a feature where a user can do a sort of single-sign on with their Twitter account, making it simple for application developers to request access to this global account. This feature is great for users, because they only have to authenticate once with the device, and still have control over which applications are allowed to tweet for them.

This feature is also great for developers. It makes tweeting really, really easy. If you have ever worked with any OAuth service on an embedded device, you know there is a bit of a disconnect between how OAuth works and native applications. In short, OAuth was designed to work within a browser and our application was not written on HTML. One solution was to pop up a UIWebView and do the authentication through it. This works, but it is clumsy. This new approach is much better.

Not only does this new Twitter API provided a simple way to authenticate, it provides a simple component where the user can enter their tweet, as shown in Figure 9–12.

Figure 9–12. *Tweeting from Belt Commander*

In Figure 9–12, we see the built-in component for tweeting on iOS. On the top we have the place where the 120 characters can be typed. To the right, there is a paperclip holding a Safari icon. This indicates that a link was attached to this tweet. The user can also add a location, cancel, or send the tweet. The code to populate this view with default values and display them is shown in Listing 9–6.

Listing 9–6. *RootViewController.m (tweetButtonClicked:)*

```
- (IBAction)tweetButtonClicked:(id)sender {
    TWTweetComposeViewController* tweetSheet = [[TWTweetComposeViewController alloc]
init];

    tweetSheet.completionHandler = ^(TWTweetComposeViewControllerResult result) {
        //view result here
    };

    [tweetSheet setInitialText:@"Check out this iOS game, Belt Commander!"];
    NSURL* url = [NSURL URLWithString:@"http://itunes.apple.com/us/app/belt-
commander/id460769032?ls=1&mt=8"];
    [tweetSheet addURL: url];

    [self presentModalViewController:tweetSheet animated:YES];
}
```

In Listing 9–6, we see the `tweetButtonClicked:` task of the class `RootViewController`. This task is called when a user clicks the Tweet button. To display the Twitter component, we simply create a new `TWTweetComposeViewController` and pass it to `presentModalViewController`. If we want to know if the tweet was sent, we can assign a block to the value `completionHandler` of `tweetSheet`. To set the initial text of the tweet, we call `setInitialText`, passing in our text. URLs and images can be added by calling `addURL:` or `addImage:`.

Of course, not every user will have Twitter set up on their device. As a developer, you want to be able to test this, which turns out to be as simple as the code in Listing 9–7.

Listing 9–7. *RootViewController.m (initTwitter)*

```
-(void)initTwitter{

    [tweetButton setEnabled:[TWTweetComposeViewController canSendTweet]];

}
```

In Listing 9–7, we see the simple task `initTwitter`, which is called when the containing `UIViewContoller` is loaded. In this task, we simply set `canSendTweet` to see if we can tweet. By using the returned value to set the enabled status of `tweetButton`, we prevent the user from being able to click the button until Twitter is enabled on the device.

The Twitter integration is great. It makes life very simple to include this feature. Let's look at including Facebook integration in our app and see how that stacks up to this built-in functionality.

Facebook Integration

It is hard to say if Twitter integration or Facebook integration is a more common request. Thankfully, Facebook has done the hard part when it comes to providing its own integration. Like Twitter, Facebook provides a set of web services that can be used to integrate with their Social Graph. But that's the hard way to do it; the easy way is to use a library called the Facebook iOS SDK, also written by Facebook.

Getting Started with the iOS and Facebook

The best place to start is to download the source code from the project `facebook-ios-sdk` from guthub. They project can be found at `https://github.com/facebook/facebook-ios-sdk`.

Once downloaded, you will find the code has been distributed as an Xcode project, making it easy to open and explore. The easiest way to get the important code into your project is to simply drag the file group named `FBConnect` from the Facebook project to your project. Figure 9–13 shows the contents of the FBConnect project.

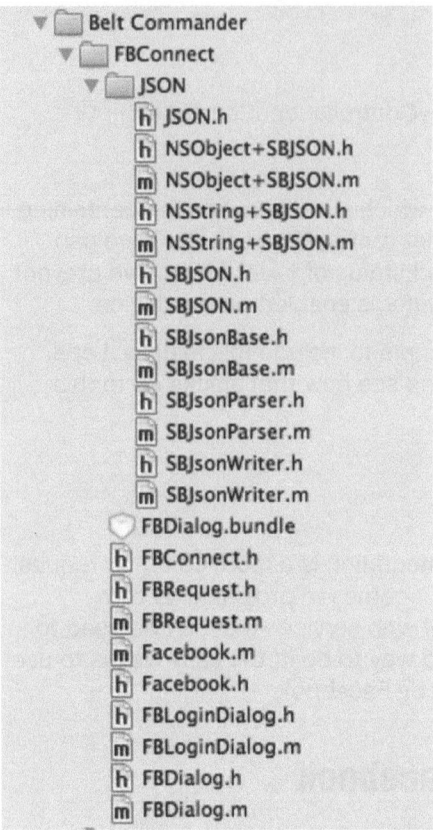

Figure 9–13. *The important classes from the Facebook iOS Xcode project*

In Figure 9–13, we see a number of files from the Facebook iOS project. The classes that start with FB are Facebook specific, while the classes in the JSON group are generic classes for dealing with JSON strings and are a dependency for the Facebook classes. The file FBConnect.h is the main header file, and will be the one you probably want to import. The class Facebook is the main class for working with Facebook. Generally, you create a singleton of this class and do all Facebook related operations through it.

Before we can look at the source code, though, we need to go to our Facebook account and create a Facebook application.

Creating a Facebook Application

The Facebook application can be the thought of as the Facebook component of our application. Facebook has us do this so there is an identity associated with any Facebook action performed on behalf of the user. Facebook also uses the Facebook application as a security context. When the user agrees to let our application make posts, they are saying that our new Facebook application is allowed to make posts.

Figure 9–14 shows the Facebook application I created for the sample game Belt Commander.

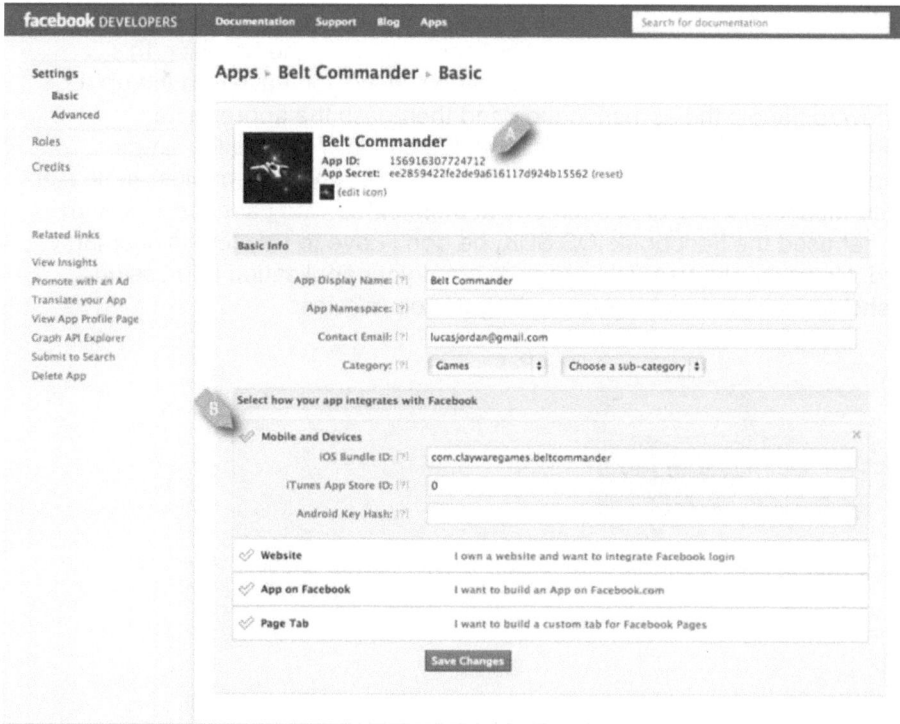

Figure 9–14. *Facebook application*

In Figure 9–14, we see the settings page for a Facebook application. At item A, you can see the App ID. We will use this value in our code when we create our Facebook object and authenticate. At Item B, you will notice that Facebook wants to you add the iOS Bundle ID. I don't know what it does with this information. When I was developing the code for this chapter, I had the wrong value in there and everything worked. Your mileage may vary. Creating a Facebook application is not hard, but I don't want to document it here, because the process seems to change every few months. Basically, you need to register your Facebook account as a developer and add a new application. Directions can be found at https://developers.facebook.com/apps.

Once you get the source code copied into your project and have created a Facebook application, you are ready to start using integrating Facebook into an application. Let's take a look at authenticating our user next.

Facebook Authentication

If you are using Facebook in a browser, you obviously authenticate through a web page that is owned by Facebook. In this way, you are trusting that the password information is going directly to Facebook and no one else. This makes a lot of sense on the web, because we you don't want other web sites handling your user's credentials. Honestly, as an app developer, I am happy to never have to worry about the user's credentials for a third-party service.

However, when it comes to developing applications outside of the browser, this authentication mechanism can be a bit of a nuisance, since you have to bring up a browser window to handle the authentication and then pass the appropriate authentication information back your application. It's a pain. To make life simpler, Facebook implements all of this for you, and it's done in a way that enables single sign on for the user. Meaning, if a user has already authenticated with Facebook through another app that used the Facebook iOS SDK, he don't have to reenter his username and password. However, the user still needs to grant your application permissions. Figure 9–15 shows Facebook asking a user to grant permissions.

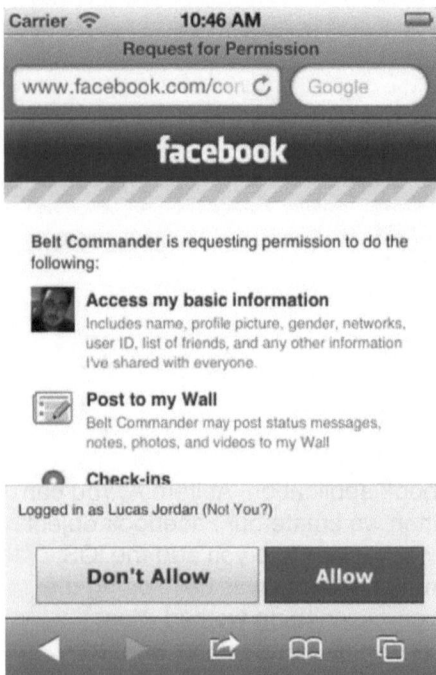

Figure 9–15. *Facebook asking a user to grant permission to an application*

In Figure 9–15, we see a Facebook page in Safari asking for the user to grant permissions to the Facebook application named Belt Commander. This web page was brought up when the iOS application Belt Commander asked to authenticate the user through Facebook. The Facebook code we included in our project will use Safari to authenticate the user if the native Facebook application is not installed; otherwise, the native one will be used. To enable Facebook to use either the native Facebook application or Safari, our application must be registered to open a URL containing our Facebook application ID.

> **NOTE:** If the application is not configured to open a URL, then Facebook will fall back to using a `UIWebView` in a popup within the application. I think the best user experience is to shoot for either the native application or Safari, by making sure your application will open the correct URL. We will take another look at this in a bit.

Initializing Facebook

Before we can authenticate we have to create and initialize our Facebook object so we can perform our authentication process. Let's take a look at how we initialize Facebook in our application, as shown in Listing 9–7.

Listing 9–7. *RootViewController.m (initFacebook)*

```
-(void)initFacebook{
    facebook = [[Facebook alloc] initWithAppId:FB_APP_ID andDelegate:self];

    NSUserDefaults* defaults = [NSUserDefaults standardUserDefaults];
    facebook.accessToken = [defaults objectForKey:@"AccessToken"];
    facebook.expirationDate = [defaults objectForKey:@"ExpirationDate"];

}
```

In Listing 9–7, we see the task `initFacebook` that is called once when the parent `UIViewController` is loaded. In this task, we create a new Facebook object by passing in an `NSString` containing our application id and specifying `self` as the delegate. After that, we pull two objects out of the standard user defaults: an access token and an expiration date. These objects, if they exist, are from a previous authentication session and will prevent the user from needing to authenticate when we call authenticate later.

The delegate object passed to `initWithAppId:andDelegate:` must conform to the protocol `FBSessionDelegate`, which defines the tasks `fbDidLogin`, `fbDidNotLogin`, and `fbDidLogout`. For this example, we only care about `fbDidLogin`, as shown in Listing 9–8.

Listing 9–8. *RootViewController.m (fbDidLogin)*

```
- (void)fbDidLogin{
    NSUserDefaults* defaults = [NSUserDefaults standardUserDefaults];
    [defaults setObject:facebook.accessToken forKey:@"AccessToken"];
    [defaults setObject:facebook.expirationDate forKey:@"ExpirationDate"];
    [defaults synchronize];
}
```

In Listing 9–8, we see the task `fbDidLogin` defined by `FBSessionDelegate` and implemented by `RootViewController`. This task is called when the user is successfully authenticated. To streamline our user experience, we record the access token and the expiration date in the standard user defaults so we can retrieve them next time the application starts, as shown in Listing 9–7.

Authentication with Facebook

Once the Facebook object is created and configured, we are ready to authenticate. In our example application, we are going to have the user click the Facebook button on the Welcome screen in order to authenticate them. In other applications, it might make sense to try and authenticate at some other time. It doesn't really matter when you authenticate, as long as you do it before you perform any other operations with the Facebook object. Listing 9–9 shows the authentication code.

Listing 9–9. *RootViewController.m (facebookButtonClicked:)*

```
- (IBAction)facebookButtonClicked:(id)sender {
    NSMutableArray* permissions = [NSMutableArray new];
    [permissions addObject: @"publish_stream"];
    [permissions addObject:@"publish_checkins"];
    [permissions addObject:@"user_about_me"];

    [facebook authorize:permissions];
}
```

In Listing 9–9, we see that code that is called the Facebook button is clicked. In this task, we create an NSMutableArray containing NSString instances for each Facebook permission we are requesting. In this case, we want to be able to publish to streams, publish check-ins, and get some basic information about the user. These are basically example permissions for this application—all we really need is publish_stream. The array of permissions is simply passed to authorize, which in turn will bring up whichever authorization mechanism is appropriate, native Facebook, Safari, or a UIWebView in a popup.

As mentioned, to support either native Facebook authentication or Safari authentication, the application must be configured to open a URL containing the Facebook application name as the protocol section of the URL. Figure 9–16 shows the first step in this configuration.

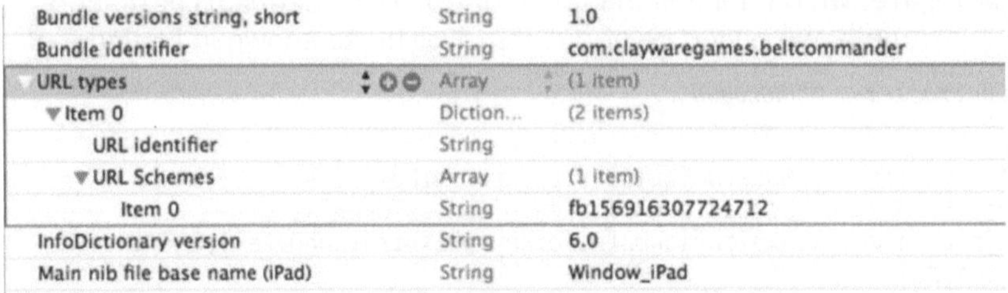

Bundle versions string, short	String	1.0
Bundle identifier	String	com.claywaregames.beltcommander
URL types	Array	(1 item)
▼ Item 0	Diction...	(2 items)
URL identifier	String	
▼ URL Schemes	Array	(1 item)
Item 0	String	fb156916307724712
InfoDictionary version	String	6.0
Main nib file base name (iPad)	String	Window_iPad

Figure 9–16. *Setting URL types to respond to a URL with a given protocol*

In Figure 9–16, we see section of the Info panel in Xcode. We have added a new row called URL types of type array. In the first item in that array, we added a dictionary with another array named URL Schemes. The first item in URL Schemes has a value of fb[your_fb_app_id], as shown with the app ID for the Facebook application Belt Commander. The next step to configure your application to properly authenticate with Facebook is to make your application delegate respond to application:handleOpenURL:, as shown in Listing 9–10.

Listing 9–10. *AppDelegate.m (application:handleOpenURL:)*

```
-(BOOL)application:(UIApplication *)application handleOpenURL:(NSURL *)url{
    UIWindow* window = [application.windows objectAtIndex:0];

    RootViewController* rvc = (RootViewController*)[window rootViewController];
    Facebook* facebook = [rvc facebook];

    return [facebook handleOpenURL:url];
}
```

In Listing 9–10, we see the task application:handleOpenURL: as defined by
AppDelegate. This task gets call when Safari or the native Facebook applications tries to
pass control back to our application. By calling handleOpenURL on our Facebook object
and pass in the URL, we complete our Facebook authentication setup. When this gets
called, fbDidLogin will be called (from Listing 9–8) and we can record our authentication
information. More important, we can now make Facebook API calls on behalf of our
user. Let's look at this next.

Facebook API Calls

Now that we have looked at setting up Facebook for authentication, we want to explore
a little about making API calls with Facebook. This is the real reason we wanted to
include Facebook in our application. Facebook provides a number of helpful tasks to
perform operations with its Social Graph API. We are not going to cover all of the details
of this AP. Facebook's excellent document can be found at:
http://developers.facebook.com/docs/reference/api/.

In our sample application we are going to post a message every time we end a game,
reporting our score. The code to do this is shown in Listing 9–11.

Listing 9–11. *RootViewController.m (notifyFacebook)*

```
-(void)notifyFacebook{
    if ([facebook isSessionValid]){
        NSString* desc = [[@"I just scored " stringByAppendingFormat:@"%i",
[beltCommanderController score]] stringByAppendingString:@" points, on Belt Commander"];

        NSString* appLink = @"http://itunes.apple.com/us/app/belt-
commander/id460769032?ls=1&mt=8";

        NSMutableDictionary* params = [NSMutableDictionary dictionaryWithObjectsAndKeys:
                                    FB_APP_ID, @"app_id",
                                    appLink, @"link",
                                    @"Presented by ClayWare Games, LLC", @"caption",
                                    desc, @"description",
                                    @"A new high score!",  @"message",
                                    nil];

        [facebook requestWithGraphPath:@"me/feed" andParams:params andHttpMethod:@"POST"
andDelegate:self];
    }
}
```

In Listing 9–11, we see the task notifyFacebook as defined by RootViewController. This task is called at the end of every game. The first thing we do is check to see if the user has authenticated with Facebook. If she has, then we assemble all of the pieces of our message in an NSMutableArray. The only item that will be included is the ID of our Facebook application with the key app_id. Notice that we are also including a link to the iTunes Store for our application; this will hopefully drive a few more downloads from our user's Facebook friends.

Once the parts of our message is all assembled, we simply call requestWithGraphPath:andParams:andHttpMethod:andDelegate:. By specifying a POST, we are indicating that we simply want to apply this to our user's feed (me/feed) without their interaction. By setting the delegate to self we will be notified if this operation succeeds (or not). In this simple example, we don't actually care if it succeeds. If all goes well, the user will see something like Figure 9–17.

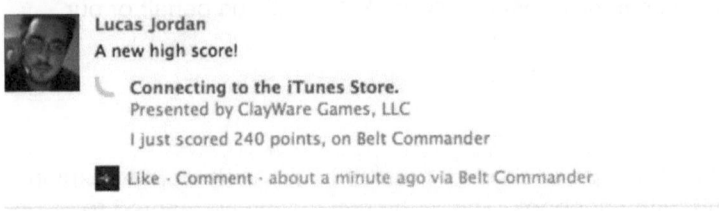

Figure 9–17. *Successful posting on Facebook*

In Figure 9–17, we see that the post to my Facebook page was successful. Like I said earlier, there is a lot more you can do with Facebook, and I encourage the reader to look into it.

As we can see the Facebook API is not as easy to use as the built-in Twitter support. But I still think it is pretty easy once you work through it the first time. It will be interesting to see if Facebook gets the same sort of treatment as Twitter did by the iOS developers at Apple.

Summary

In this chapter, we explored Game Center and introduced what services it provides to the user. We explored those features in terms of GameKit to see how we can post the user's score to a leaderboard and award achievements. Continuing with the social trend, we looked at the built-in support for Twitter in iOS 5. Lastly, we looked at integrating Facebook into an iOS application. We covered how to handle authentication as well post a simple message to the users feed.

Chapter 10

Monetizing via the Apple App Store

There are three main ways to make money by having apps in the Apple App Store. The most obvious way is to simply charge for the application. The second is to give the app away for free, but include ads. Another way, which can work for free and paid apps, is to include things for sale within the app. These are called in-app purchases, and are very popular with some styles of games.

This chapter will not cover what makes a good in-app purchase or whether your application should be free or available for a fee. It will cover the basics of setting up in-app purchases and explain the different types of purchases available.

In-App Purchases

In-app purchases are virtual items or services found within an app that are available for a fee. Many different applications take advantage of in-app purchases to make money—not just games. Some non-game examples include downloading articles or comics, cooking recipes, and detailed map data. Some in-game examples include additional levels, new powers, extra game money, and novelty items for customizing the look of the game.

We will be continuing with our example code from the game Belt Commander. In Figure 10–1, you can see the in-app purchase options for Belt Commander.

In Figure 10–1, we see a screenshot from the game Belt Commander. There are three buttons, each specifying which type of actors should be present in the game when it is played. The game comes with asteroids available by default; the two buttons on the right are used to purchase additional types of actors in the game. The button in the middle is for saucers, which were already purchased but won't be included in the next game (Inactive). The button to the right indicates that the user may choose to purchase the ability to include power-up actors in the game. The user initiates the purchase by clicking the button and walking through the purchase dialogs.

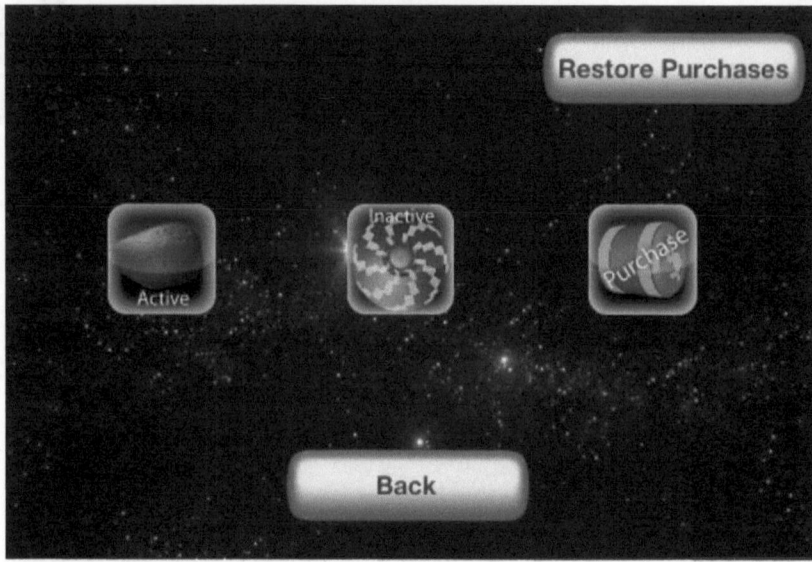

Figure 10–1. *In-app purchases in Belt Commander*

Before we implement these features, we will look at the different types of in-app purchases available and how to enable in-app purchases in our application.

Overview of Purchase Types

There are four distinct types of in-app purchases, each supporting a different business need. If these four types of in-app purchases do not fit your business needs, you will have to roll your own. Writing your own in-app purchase service could be very time consuming. I would recommend seeing if your business needs can be adapted to fit Apple's model. The benefit is not only a decrease in development time, but also there is a benefit the user. When you use Apple's system, your user will see a familiar interface and login with a familiar set of credentials: their Apple ID. The four types of in-app purchases are as follows:

- Non-consumable
- Consumable
- Subscriptions
- Auto-renewing subscriptions

Let's take a closer look at each of the four types of in-app purchases.

Non-consumable

Non-consumable in-app purchases are items that are purchases once and grant persistent access to some item or service. Non-consumable items can be re-downloaded by the user, and should be available on all of the user's devices. Non-consumable purchase can include additional game levels, new types of enemies to fight, or silly hats to customize your user's game character.

Consumable

Consumable in-app purchases are designed to support items or services that are consumed. In a game, this might include additional lives, points for buying in-game items, or other things that can be used up. Consumable purchases are designed to be repeated.

Consumable purchases cannot be downloaded again, so by their nature they will not be available across a user's devices unless you implement your own server to share this information across accounts.

Subscriptions

Subscriptions are purchases with a time limit. After a given amount of time, the user will no longer have access to the purchase. These types of purchases might include access to articles, location information, or some other valuable set data. Subscription purchases cannot be re-downloaded and must be repurchased to get the same service after a device restore (if backup fails) or on a second iOS devices owned by the user.

Auto-Renewing Subscriptions

Auto-renewing subscriptions are purchases for items or services for a duration of time, much like regular subscriptions. The difference is that auto-renewing subscriptions are re-downloadable. They represent access to a service that is associated with the user, not the device. The user can access this purchase on any iOS devices they own.

Now let's look at what is required to ready an application for in-app purchases.

Preparing for In-app Purchases

Much like Game Center, an application must be enabled in iTunes Connect to include in-app purchases. The requirements are exactly the same: you have to use an App ID that does not have a wildcard. In Chapter 9, we cover the process of creating an App ID in your provisioning center that works with Game Center. You can follow the directions to create a second App ID, or simply add in-app purchases to that same game.

Enabling and Creating In-App Purchases

Once you have an app in iTunes Connect, enabling in-app purchases is done by pressing the Manage In-App Purchases button, as shown in Figure 10–2.

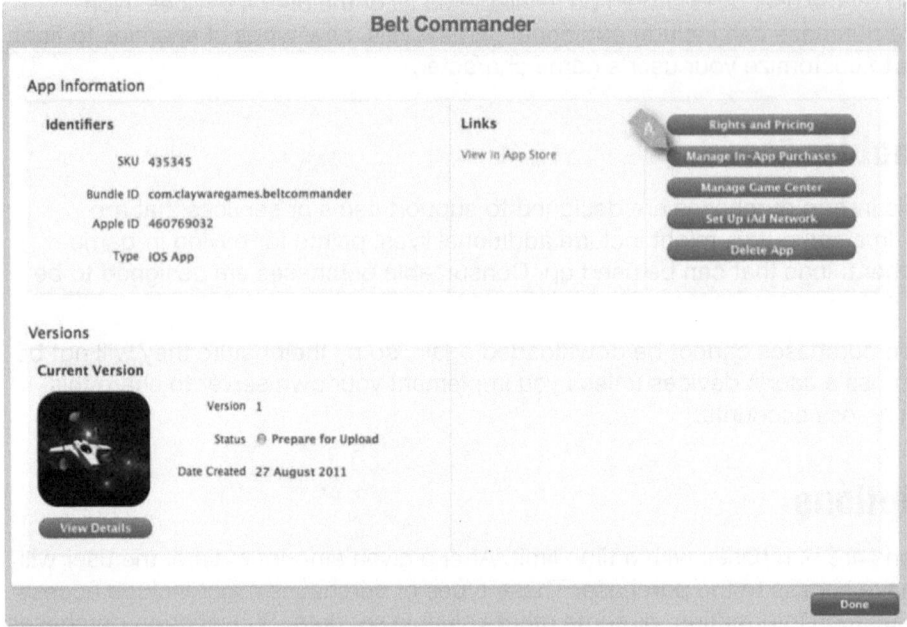

Figure 10–2. *Enabling In-app Purchases in iTunes Connect*

In Figure 10–2, we see the view in iTunes Connect for configuring an application. Item A shows the button to click to enable in-app purchases. After clicking this button, you will be able to add items to you app that users may purchase. Figure 10–3 shows the items I prepared for Belt Commander.

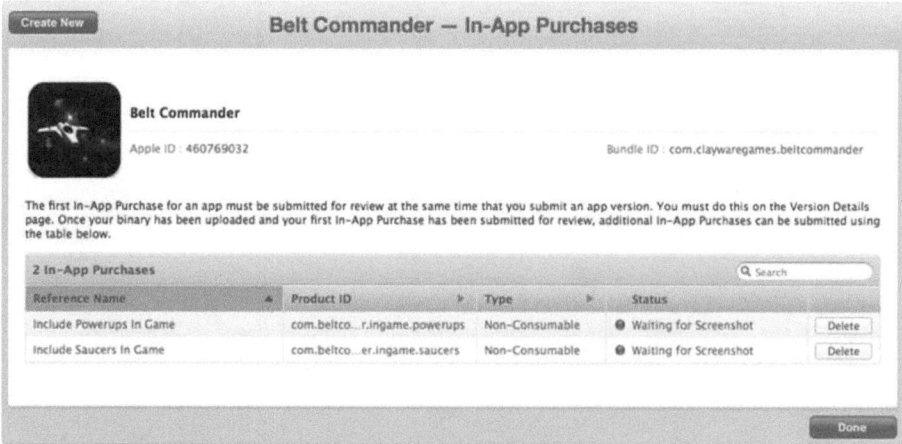

Figure 10–3. *In-App purchases available for Belt Commander*

In Figure 10–3, we see the screen in iTunes Connect where in-app purchases are listed. We can see that we added two: one for each additional type of actor the user can add to the game. For each in-app purchase, we see a reference name, a product ID, a product type, and the product status. The reference name is for internal use—it will not be seen by the user. The product ID is a unique name, used across all iOS applications, that identifies the purchase. The product ID will be how we reference this purchase in code.

The type refers to the different purchase types we discussed earlier—in this case, non-consumable. The status value indicates what is required for this in-app purchase to be accepted by Apple. Apple must review each in-app purchase you create, much like how each app must be reviewed before sale. In this case, we are informed that Apple is waiting for screenshots of the purchase before it will review it. The screenshots are not shown to the user—they just give context to Apple about what the in-app purchase is. In Figure 10–4, we see the view in iTunes Connect that is used to create a new in-app purchase.

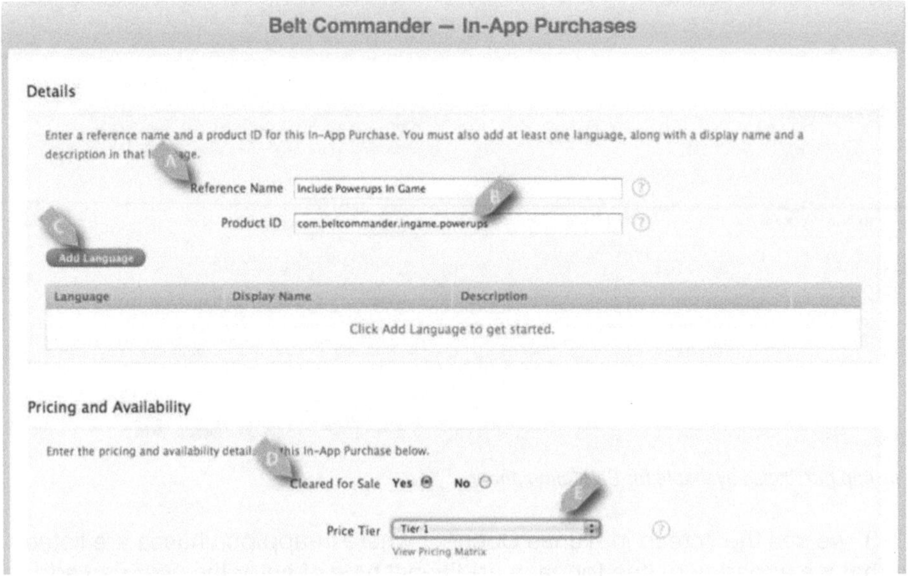

Figure 10–4. *A new in-app purchase*

Item A in Figure 10–4 shows where we enter our reference name for the in-app purchase. Item B shows the Product ID for the in-app purchase. In order for you to create the in-app purchase, you must provide the localized strings for the in-app purchase by clicking the Add Language button (item C).

Under the Pricing and Availability section of Figure 10–4, you must set the item as cleared for sale, at item D, for you to test the purchase. You must also specify a Price Tier at item E. The price tiers are the same as the tiers for an application, with the exception that there is no free tier.

Once you have in-app purchases defined for you application, you must configure a test user so you can test your purchases. It is highly recommended that you test your in-app purchases before you submit your application.

Creating a Test User

Once you have in-app purchases defined, you will want to create a test user to "purchase" those items in your application. This give you a chance to test enabling or disabling each purchased feature with the same workflow that a real user will experience. Figure 10–5 shows where you start the process of adding a test user in iTunes Connect.

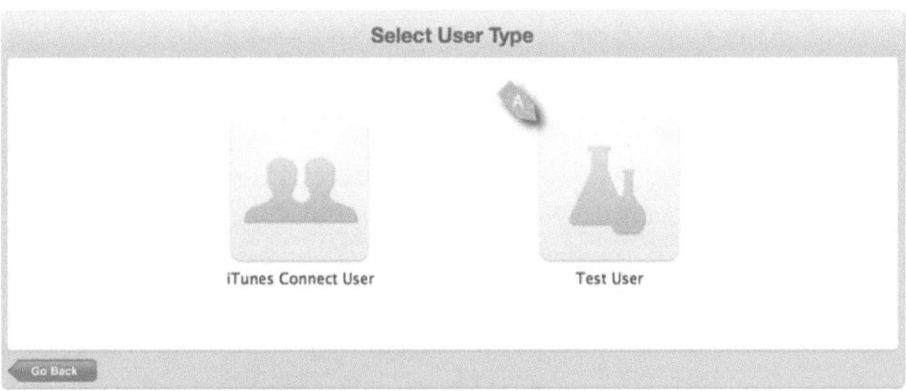

Figure 10–5. *Creating a test user*

In Figure 10–5, we see the option to create a Test User from the Select User Type view. This view is available in iTunes Connect by clicking on the Manage Users button found on the home screen.

> **NOTE:** I am emphasizing the point about making sure you use a test user to debug your in-app purchases because trying to debug with a real user will confound your efforts.

Once you click the Test User button shown in Figure 10–5, you will be prompted to create a new user. There should be nothing surprising—just keep in mind that you should not use an e-mail address that is currently an Apple ID.

We have looked at adding in-app purchases to our application in iTunes Connect and how to make sure we have a test user to debug our application. Next, we will look how in-app purchases are integrated into Belt Commander, giving us an understanding of the classes involved and how to use them.

Class and Code for In-App Purchases

Creating in-app purchases in iTunes Connect is just the first step in actually having in-app purchases available in your application. The next step is to understand the process of how in-app purchases are actually purchased and what classes are involved. The framework that contains the pertinent classes is called StoreKit, and is available as part of you iOS SDK installation. Using these classes, we are going to implement the workflow as shown in Figure 10–6, which shows the process of determining which products are available and how to handle the request to make a purchase.

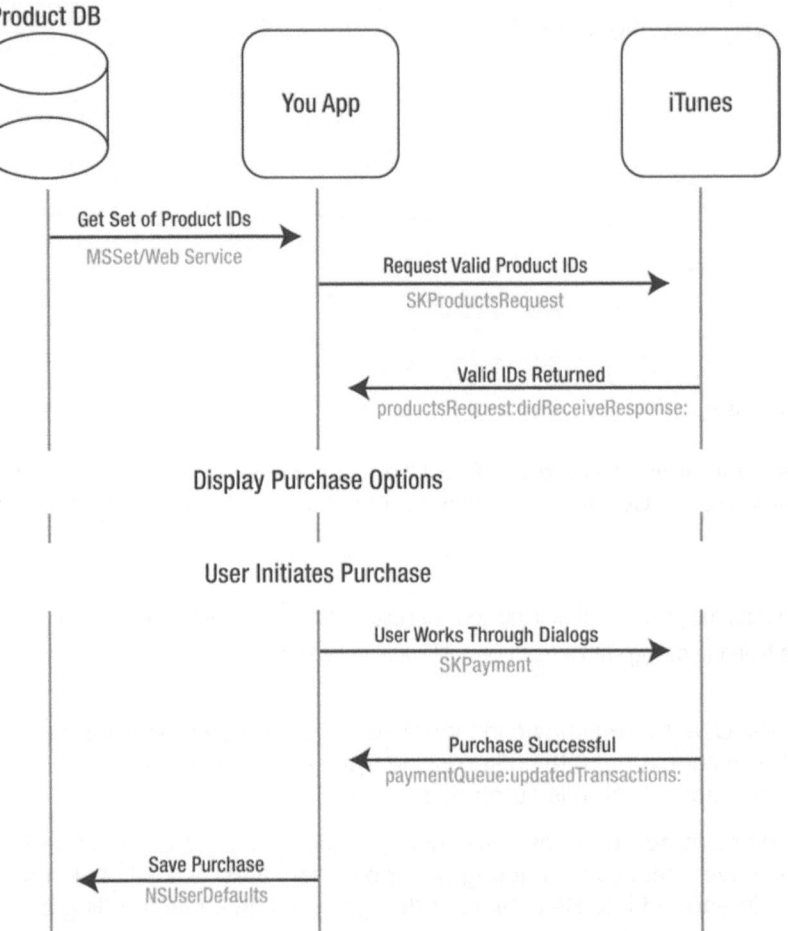

Figure 10–6. *Purchase Process*

Starting at the top-left of Figure 10–6, we begin by getting a set of Product IDs for products we want to present to the user. This list might come from an external web service or just be a hardwired set of strings in the application. How you get this list will depend on how you are managing what is for sale in your app. If you are adding new products all the time, you probably want to pull the list of products from a web service. If you are just getting started with in-app purchases, a fixed list might work for you, but you will have to update your app to change the list. Once we have our list of possible Product IDs, make a request to the iTunes Store, asking which of the IDs we have passed along are valid and ready for sale. The IDs returned from the iTunes Store represent the items we want to make available to the user. The iTunes Store will not return Product IDs that are unknown or not ready for sale. This insures we are presenting only valid items that have gone through Apple's review process. In addition to letting us know that the IDs are valid, the iTunes Store also returns the localized strings associated with the purchase, allowing us to present language-appropriate text to the user that matches exactly what was entered into iTunes Connect.

Continuing with the workflow shown in Figure 10–6, at some point the user will click a button to purchase an item. When this happens, we simply create a payment request and submit it, at which point the user will be prompted with a number of dialogs confirming the purchase. If all goes well, we will get a callback letting us know that purchase was successful. When this happens, we need to record that the purchase has been made. We can do this simply by sticking the Product ID in the standard NSUserDefaults, or we can do something more sophisticated, like save it to a web service. The dialogs that the user must work through in order to confirm a purchase are shown in Figure 10–7.

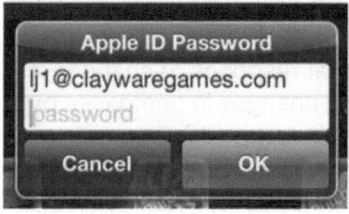

Figure 10–7. *Making a purchase*

The dialog on the upper left side of the figure will always be shown, as well as the one the lower right. The other two dialogs will be shown between the first and last dialog if the user has to log in with his Apple ID. It is important to remember that, as a developer, we have no control over exactly which dialogs will be displayed or how long the user will take to complete the process. It is important not to expect the user to complete this process while the action of the game is in process.

Now let's look at a specific example, where we implement the purchase of a non-consumable in-app purchase for our game Belt Commander. We are going to implement the view from Figure 10–1, where the user can select which actors are involved in the game and purchase new actors to be made available.

In-App Purchase Implementation

As we discussed, the process of making a purchase is really just two calls to the iTunes Store. We will see in this section that it really only involves a handful of classes. In our example, we will be handling most of the details of working with StoreKit and handling the UI in a single class called ExtrasController, whose header is shown in Listing 10–1.

Listing 10–1. *ExtrasController.h*

```
@interface ExtrasController : UIViewController<SKPaymentTransactionObserver,
SKProductsRequestDelegate>{
    GameParameters* gameParams;
}
@property (strong, nonatomic) IBOutlet UIButton *asteroidButton;
@property (strong, nonatomic) IBOutlet UIButton *saucerButton;
@property (strong, nonatomic) IBOutlet UIButton *powerupButton;

-(void)setGameParams:(GameParameters*)params;
@end
```

In Listing 10–1, we see the header for the class ExtrasController. The class ExtrasController plays two roles: it manages the purchasing process and updates the three buttons accordingly. The object gameParams is responsible for storing which actors are enabled in the game and which purchases have been made. In order to interact with StoreKit, ExtrasController conforms to the protocols SKPaymentTransactionObserver and SKProductsRequestDelegate. These protocols provide everything you need for your application to respond to interactions with the store. We will see how ExtrasController implements the tasks from these two protocols as we work through the code.

The first step when working StoreKit is to set up an object that responds to changes in the state of purchases. We create an object that responds to changes in purchases by adding an observer to the default SKPaymentQueue, as shown in Listing 10–2.

Listing 10–2. *ExtrasController.m (viewDidLoad)*

```
#define PURCHASE_INGAME_SAUCERS @"com.beltcommander.ingame.saucers"
#define PURCHASE_INGAME_POWERUPS @"com.beltcommander.ingame.powerups"
//….
- (void)viewDidLoad
{
    [super viewDidLoad];
    [[SKPaymentQueue defaultQueue] addTransactionObserver:self];

    //Get set of product ids
    NSSet* potentialProcucts = [NSSet setWithObjects:PURCHASE_INGAME_POWERUPS,
PURCHASE_INGAME_SAUCERS, nil];

    SKProductsRequest* request = [[SKProductsRequest alloc]
initWithProductIdentifiers:potentialProcucts];
    [request setDelegate:self];
    [request start];

    [self setGameParams: [GameParameters readFromDefaults]];
}
```

In Listing 10–2, we see the task viewDidLoad for the class ExtrasController. In this task, we start by simply getting the default SKPaymentQueue by calling defaultQueue, adding self as an observer. Because self conforms to the protocol SKPaymentTransactionObserver, our ExtrasController will be informed of all changes in state for a purchase. Once we have a registered observer, we create a set of known product IDs. In this case, the NSSet potentialProducts is created from two constants: PURCHASE_INGAME_POWERUPS and PURCHASE_INGAME_SAUCERS. As we can see, these constants are the same strings that we specified when we created the products in iTunes Connect.

Once we have specified our set of potential products we create an SKProductsRequest, set self as the delegate, and call start on it. While StoreKit is working out which Product IDs are valid, we call setGameParams and specify that we should use the GameParameters object we have stored the default NSUsersDefaults. When StoreKit has its results, the tasks productsRequest:didReceiveResponse:, as shown in Listing 10–3.

Listing 10–3. *ExtrasController.m (productsRequest:didRecieveResponse:)*

```
- (void)productsRequest:(SKProductsRequest *)request
didReceiveResponse:(SKProductsResponse *)response {
    for (SKProduct* aProduct in response.products){
        if ([aProduct.productIdentifier isEqualToString:PURCHASE_INGAME_SAUCERS]){
            [saucerButton setEnabled:YES];
            [saucerButton setHidden:NO];
        }
        if ([aProduct.productIdentifier isEqualToString:PURCHASE_INGAME_POWERUPS]){
            [powerupButton setEnabled:YES];
            [powerupButton setHidden:NO];
        }
    }
}
```

In Listing 10–3, we see the task productsRequest:didReceiveResponse: that gets called when we have a valid set of Product IDs. In this task, we simply iterate over all of the products returned and make our buttons visible for purchasing saucers and power-ups. In this way, we make sure the user only sees viable purchase options.

We have looked at how we figure out which purchase are valid. Before we look at how purchases are actually made, we should look at the class GameParameters. The class GameParameters is responsible for storing which products have already been purchased and which actors the users wants in their game.

Driving the UI from Existing Purchases

The three buttons, shown in Figure 10–1, allow the user to select which actors are available in the game. The same UI allows the user to purchase new actors. In order to keep track of this information we have created the model class GameParameters, whose header is shown in Listing 10–4.

Listing 10–4. *GameParameters.h*

```
@interface GameParameters : NSObject<NSCoding>

@property (nonatomic) BOOL includeAsteroids;
@property (nonatomic) BOOL includeSaucers;
@property (nonatomic) BOOL includePowerups;
@property (nonatomic, retain) NSMutableSet* purchases;

+(id)gameParameters;
+(id)readFromDefaults;
-(void)writeToDefaults;
@end
```

In Listing 10–4, we see the header for the class GameParameters. We have three BOOL properties to keep track of which actors the user wants in the game. We also have an NSMutableSet called purchases that is used to store a set of product ID strings. The constructor gameParameters will create a new GameParameters object with no purchases and only includeAsteroids set to true (implementation not shown). The class task readFromDefaults is used to pull a GameParameter out of the default NSUserDefaults whereas writeToDefaults is used to serialize a GameParameter back to the default NSUserDefautls. Listing 10–5 shows both readFromDefaults and writeToDeafaults.

Listing 10–5. *GameParameters.m (readFromDefaults and writeToDefaults)*

```
+(id)readFromDefaults{
    NSUserDefaults* defaults = [NSUserDefaults standardUserDefaults];
    NSData* data = [defaults objectForKey:GAME_PARAM_KEY];
    if (data == nil){
        GameParameters* params = [GameParameters gameParameters];
        [params writeToDefaults];
        return params;
    } else{
        return [NSKeyedUnarchiver unarchiveObjectWithData: data];
    }
}
-(void)writeToDefaults{
    NSData* data = [NSKeyedArchiver archivedDataWithRootObject:self];
    NSUserDefaults* defaults = [NSUserDefaults standardUserDefaults];
    [defaults setObject:data forKey:GAME_PARAM_KEY];
    [defaults synchronize];
}
```

In Listing 10–5, we see the tasks readFromDefaults and writeToDefaults. These two tasks are opposites: readFromDefaults will return a GameParameter for the key GAME_PARAM_KEY as stored in the user defaults. The task writeToDefaults writes a GameParameter to the user defaults for the key GAME_PARAM_KEY. They key idea to both of these tasks is that we store a GameParameter in the user defaults to remember the user's purchases by recording the keys in the property purchases. The class GameParameters can be stored in an NSUserDefautls, because it implements the tasks from the protocol NSCoder. See Chapter 2 for a description of that process.

Let's take a look at how we use this class is used to drive the UI and set the states of the three different buttons from Figure 10–1. This is handled in the task setGameParameters:, as shown in Listing 10–6.

Listing 10–6. *GameParameters.m (setGameParameters:)*

```
-(void)setGameParams:(GameParameters*)params{
    gameParams = params;
    NSSet* purchases = [gameParams purchases];

    if ([params includeAsteroids]){
        [asteroidButton setImage:[UIImage imageNamed:@"asteroid_active"]
forState:UIControlStateNormal];
    } else {
        [asteroidButton setImage:[UIImage imageNamed:@"asteroid_inactive"]
forState:UIControlStateNormal];
    }

    if ([purchases containsObject:PURCHASE_INGAME_SAUCERS]){
        if ([params includeSaucers]){
            [saucerButton setImage:[UIImage imageNamed:@"saucer_active"]
forState:UIControlStateNormal];
        } else {
            [saucerButton setImage:[UIImage imageNamed:@"saucer_inactive"]
forState:UIControlStateNormal];
        }
    } else {
        [saucerButton setImage:[UIImage imageNamed:@"saucer_purchase"]
forState:UIControlStateNormal];
    }

    if ([purchases containsObject:PURCHASE_INGAME_POWERUPS]){
        if ([params includePowerups]){
            [powerupButton setImage:[UIImage imageNamed:@"powerup_active"]
forState:UIControlStateNormal];
        } else {
            [powerupButton setImage:[UIImage imageNamed:@"powerup_inactive"]
forState:UIControlStateNormal];
        }
    } else {
        [powerupButton setImage:[UIImage imageNamed:@"powerup_purchase"]
forState:UIControlStateNormal];
    }
}
```

In Listing 10–6, we see the task setGameParameters. This task is responsible for making the UI reflect the state of the passed in GameParameters object called param. For the asteroidButton, we simply indicate which image should be displayed depending on the value of the property includeAsteroids. For saucerButton, we perform a similar test, but only if the product ID for the saucers is included in the NSSet purchases. If the ID is not present, we set the image on the button to display the purchase version. We give the UIButton powerButton a similar treatment, by setting the background to be active, inactive, or purchase.

Making the Purchase

Now that we know how the button state is managed, let's look at how clicking one of the purchase buttons triggers a purchase. Listing 10–7 shows the code that is run when a user presses the saucer button.

Listing 10–7. *ExtrasController.m (saucerButtonClicked:)*

```
- (IBAction)saucerButtonClicked:(id)sender {
    if ([gameParams.purchases containsObject:PURCHASE_INGAME_SAUCERS]){
        [gameParams setIncludeSaucers:![gameParams includeSaucers]];
        [self setGameParams:gameParams];
        [gameParams writeToDefaults];
    } else {
        SKPayment* payRequest = [SKPayment paymentWithProductIdentifier:
PURCHASE_INGAME_SAUCERS];
        [[SKPaymentQueue defaultQueue] addPayment:payRequest];
    }
}
```

In Listing 10–7, we see the task saucerButtonClicked that is called when the user clicks the saucer button. If we have already purchased the saucer feature, we know that the user is simply looking to toggle if saucers are included. We respond to this by flipping the value of includeSaucers, updating the UI by calling setGameParams, and writing the change to disk. If the user has not yet purchased the saucer feature, we know she wants to make a purchase. To start this process, we simply create an SKPayment object with the appropriate product ID and add the SKPayment object to the default SKPaymentQueue. This triggers the dialog process, as shown in Figure 10–7. The code to manage the power-ups is almost identical to the code in Listing 10–7, and it's not worth looking at. Next, let's look at how we respond to a successful purchase.

Responding to a Successful Purchase

In order to respond to a successful purchase, we had to add an instance of SKPaymentTransactionObserver to the default SKPaymentQueue back in Listing 10–2. We are now ready to look at the code that gets called when the SKPaymentQueue successfully processes a purchase, as shown in Listing 10–8.

Listing 10–8. *ExtrasController.m (paymentQueue: updatedTransactions:)*

```
- (void)paymentQueue:(SKPaymentQueue *)queue updatedTransactions:(NSArray
*)transactions{
    for (SKPaymentTransaction* transaction in transactions){
        if (transaction.transactionState == SKPaymentTransactionStatePurchased ||
transaction.transactionState == SKPaymentTransactionStateRestored){
            NSString* productIdentifier = transaction.payment.productIdentifier;

            [[gameParams purchases] addObject: productIdentifier];
            [gameParams writeToDefaults];

            [self setGameParams:gameParams];
```

```
            [queue finishTransaction:transaction];
        }
    }
}
```

In Listing 10–8, we see the tasks paymentQueue:updatedTransactions: as implemented by the class ExtrasController. In this task, we iterate through each SKPaymentTransaction in the array transactions. For each SKPaymentTransaction we check to see if the state is SKPaymentTransactionStatePurchase or SKPaymentTransactionStateRestore. If it is, we add the product ID of the transaction to our gamePagam's purchases property and save it. We also update the UI by calling setGameParams:. Lastly, we inform the queue that we are done working with the transaction by calling finishTransaction:.

Technically speaking, if the task paymentQueue:updatedTransactions: as a result of the user making a new purchase the state of the transaction would be SKPaymentTransactionStatePurchased. But because the type or products we are using are non-consumable, we need to make sure the user has a way to restore his purchases if his device is reset or if he made the purchase on another iOS device. In order to handle this, we make sure we allow the transaction state to be SKPaymentTransactionStateRestored and treat it the same as a new purchase. In order to allow the user to restore we provide a restore purchases button that calls the code in Listing 10–9.

Listing 10–9. ExtrasController.m (restorePurchasesClicked:)

```
- (IBAction)restorePurchasesClicked:(id)sender {
    [[SKPaymentQueue defaultQueue] restoreCompletedTransactions];
}
```

In Listing 10–9, we see the task trestorePurchasesClicked that gets called when the user clicks the restore purchases button. In this task, we simply call restoreCompletedTransactions on the default SKPaymentQueue. This causes StoreKit to contact the iTunes Store and re-download all product IDs for the current app and user. When it has done that, it calls paymentQueue:updatedTransaction:, as shown in Listing 10–8. This is all that is required to allow the user to get access to her purchases.

Summary

In this chapter, we reviewed what in-app purchases are and the four different types. We also looked at how we implemented the purchasing of in-app products in our Belt Commander game. We reviewed how to get a valid list of product IDs from the iTunes Store and how to create a UI based on that list. We also looked at how to trigger a purchase process and how to respond to a successful completion, keeping in mind how to store a record of that purchase for the user.

Chapter 11

A Completed View Belt Commander

Throughout this book, we have covered a number of techniques for building iOS games. We looked at how to build the basic application flow for a game in Chapter 3. We learned how to build a basic turn-by-turn game when we looked at Coin Sorter in Chapter 4. Then we moved in to frame-by-frame games and learned how to build a game that is in constant motion in Chapter 5. In Chapters 6 and 7, we looked at how to create different types of actors to populate our game. Chapter 8 covered how to capture user input that manipulates in-game elements, and Chapter 9 looked at how to reach out to our players with Game Center and other social media services. In Chapter 10, we discussed how to add in app purchases and to help make some money with our game. In this chapter, we are going to bring all of these elements together to create single game, Belt Commander. The title graphic is shown in Figure 11–1.

Figure 11–1. *Title graphic for Belt Commander*

In this chapter, we are going to take a high-level look at the game Belt Commander. We are also going to walk through how the game is put together, including how the views are organized and how the application game logic is implemented. We will discuss the different actors in the game and how they are added. In some ways, this chapter is a review of the techniques we learned in previous chapters. However, exploring how these techniques fit together will give you a roadmap that will help you put together your own games. In fact, feel free to start out with the sample code for this chapter for your own game. It is my intention to do just that and release a commercial version of Belt Commander in the iTunes Store. So let's get started.

Belt Commander: Game Recap

Belt Commander is an action game in which you control a spaceship that is traveling through an asteroid belt. Destroying asteroids and alien flying saucers rewards the player with points. During play, a power-up occasionally floats onto the screen, giving the player a chance to heal some damage, earn some bonus points, or upgrade weapons. This is a relatively fast-paced game with simple interactions. Let's take a look at the opening screen of the game, shown in Figure 11–2.

Figure 11–2. *Belt Commander's opening screen*

In Figure 11–2, we see the first screen of the game. We have a title at the top of the screen as well as a number of buttons. The Facebook, Tweet, and Leaderboards buttons enable the various social functions that we discussed in Chapter 9. The extras button allows the user to configure the game and purchase new parts of the game. How purchases are handled is described in Chapter 10. For this chapter, we will assume all of the extras are purchased. The Play button will take the user into the action part of the game, as shown in Figure 11–3.

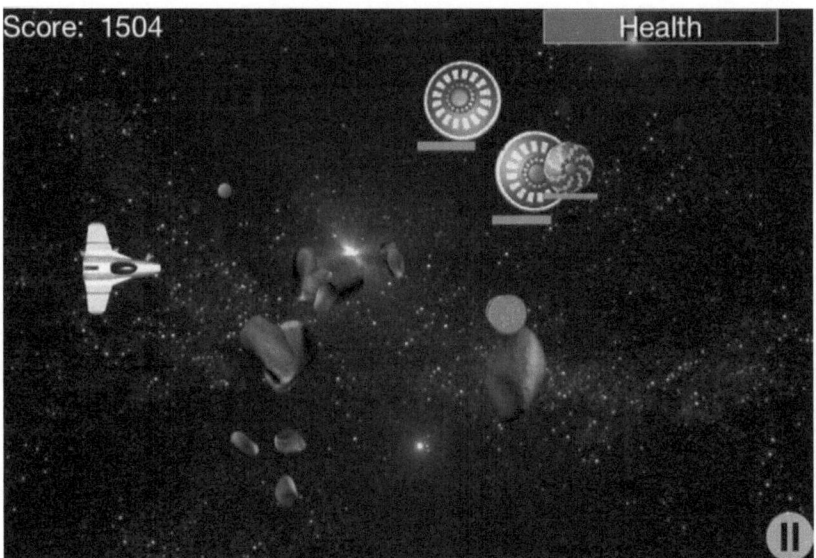

Figure 11–3. *Belt Commander in action*

In Figure 11–3, we see the action part of the game. On the left is the spaceship controlled by the user. The user can tap on the screen and the ship will travel up or down to the location of the tap. This is how the user can avoid the many asteroids that travel from the right side of the screen to the left. In addition to the asteroids, there are alien saucers that travel up and down on the screen shooting bullets at the user's ship. Both the asteroids and the saucer's bullets do damage to the ship when they hit. The game continues until the ship is overwhelmed with damage. To help the player along, power-ups appear on the right and travel to the left. If the user can intercept a power-up, he receives whatever bonus it gives: health, more damage, or extra points.

In the lower right-hand side of the Figure 11–3, we see a pause button (II). This will pause the game and allow the user to quit or continue playing, as shown in Figure 11–4.

Figure 11–4. *Belt Commander on pause*

In Figure 11–4, we see the dialog that pops up when the user clicks the touch button on the lower right of the screen. This dialog allows the user to exit the game by clicking Yes, or to continue by clicking No. If the user is playing the game and the application is sent to the background by another app starting, this dialog will be presented when the user brings the game back up. When the user has no health left, the game is over. The dialog shown in Figure 11–5 appears.

Figure 11–5. *Game over*

In Figure 11–4, we see the dialog presented at the end of the game. This dialog allows the user to get right back into the action by clicking Yes. If the user clicks No, she is sent back to the starting screen of the game. The entire application flow is shown in Figure 11–6.

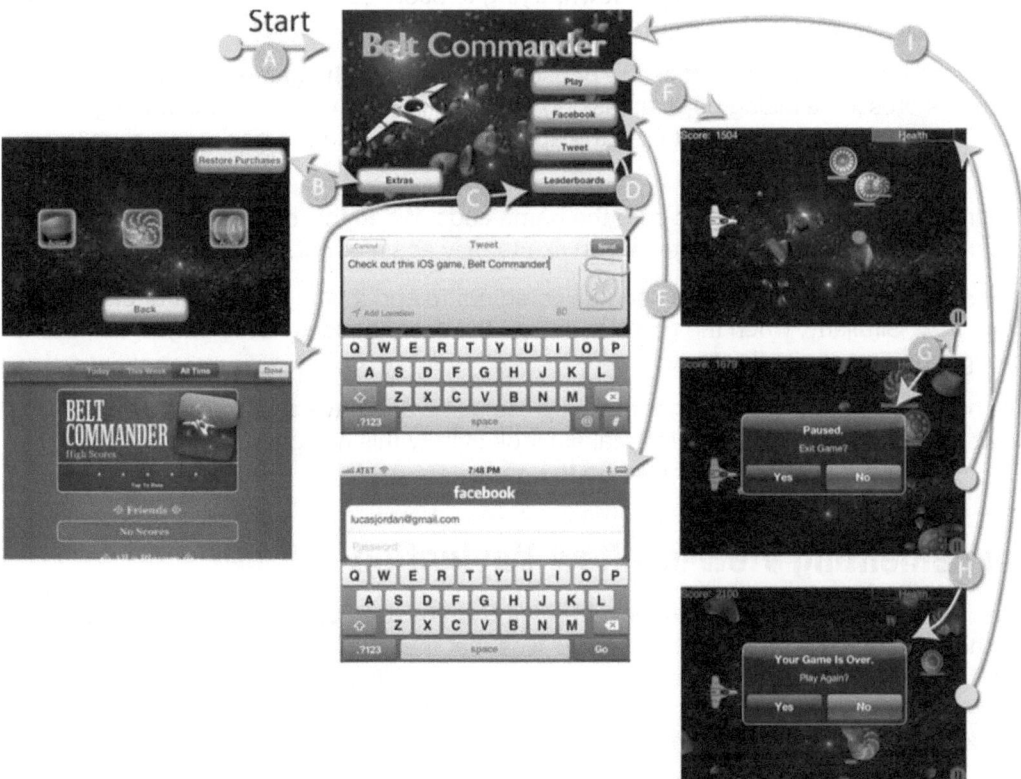

Figure 11–6. *Application flow*

In Figure 11–6, we see the eight screens of the game. The arrows indicate how the user navigates from one view to another. Each transition is lettered as described in the following:

The application starts at A, the start view. This is the first thing the user sees.

By pressing the Extras button, the user is taken to the extras view, where they can select which actors are in the game: asteroids, saucers, and power-up. They return to the start view by clicking the Back button.

Clicking the Leaderboard button brings up the standard leaderboard view. Clicking Done on the upper right returns the user to the start view.

The Tweet button brings up the Twitter dialog. While not technically a separate view, since the Tweet dialog is a pop-up, it still takes up most of the screen. The Send or Cancel button returns the user to the start view.

Clicking the Facebook button will exit the app and bring up a Facebook authentication screen. When the user enters his credentials, he is brought back to the start view.

Clicking the Play button starts the game proper. In the action view, the user plays the game. They move the ship up and down, trying to score as many points as possible by destroying asteroids and saucers.

While playing, the user may click the pause button to stop the game. When the pause button is pressed a dialog pops up asking the user if they would like to exit the game or not. This dialog is also brought up if the application is sent to the background while the game is in progress.

When the ships health reaches zero the game ends. When this happens a dialog is presented asking the user if they would like to play again or not.

If the user has paused the game and selected not to end it, she is brought back to the start view. Similarly, when the game is over, if the user chooses not to play again, she is brought back to the main view.

We have looked at the views that make up the game and have a sense of how the game flows from view to view. Next, let's take a look at how this navigation is implemented, starting with the XIB files that make up the game.

Implementing View-to-View Navigation

We have gone over the views involved in Belt Commander. Let's take a closer look at the how the game is initialized, at the XIB files that make up the game, and how we handle navigation programmatically. This information will help you when you are building your own game, as there are a number of different techniques at work here. Let's start by looking at how we launch the application.

Launching the Application

Like all iOS applications, the starting place is the main function found in the file main.m. However, in this application, we don't customize that at all, so we can skip describing it. The real starting place for this application is the plist file, Belt Commander-Info.plist. Figure 11–7 shows Xcode's representation of this file.

Key	Type	Value
Localization native development region	String	en
Bundle display name	String	${PRODUCT_NAME}
Executable file	String	${EXECUTABLE_NAME}
▶ Icon files	Array	(0 items)
Bundle identifier	String	com.claywaregames.beltcommander
InfoDictionary version	String	6.0
Bundle name	String	${PRODUCT_NAME}
Bundle OS Type code	String	APPL
Bundle versions string, short	String	1.0
Bundle creator OS Type code	String	????
▶ URL types	Array	(1 item)
Bundle version	String	1.0
Application requires iPhone environme	Boolean	YES
Main nib file base name (iPad)	String	Window_iPad
Main nib file base name (iPhone)	String	Window_iPhone
Status bar is initially hidden	Boolean	YES
▶ Supported interface orientations	Array	(3 items)
▶ Supported interface orientations (iPad)	Array	(4 items)

Figure 11–7. *Belt Commander-Info.plist in Xcode*

In Figure 11–7, we see file Belt Commander-Info.plist as presented by Xcode. In this file, we see a pretty standard setup for an iOS application. We see that the Bundle identifier is com.claywaregames.beltcommander that corresponds to the application we created in iTunes Connect that supports in-app purchases and Game Center. What is a little different in this file from the other samples in this book is that we are specifying the starting XIB files in this file. The starting XIB files are indicated by the key values "Main nib file base name (iPad)" and "Main nib file base name (iPhone)." These keys are set to the values "Window_iPad" and "Window_iPhone," respectively. By setting these values, we are requesting that the iOS application automatically use these files to create our main UIWindow and call makeKeyAndVisible on it. This removes the need for us to put any code in our application delegate's application:didFinishLaunchingWithOptions: task. To understand how the other objects in the application relate to each other, let's take a look at XIB files for this game.

The XIB Files

The XIB files in this application bind instances of the key classes together and allow them to coordinate. While this application is a universal one and runs on both the iPhone and iPad, we are just going to focus on the iPhone XIB file, because the iPad version is different only in layout. By looking at the contents of this XIB file, we will understand how the application is put together. An overview is shown in Figure 11–8.

Figure 11–8. *Overview of iPhone XIB file*

In Figure 11–8, we see the XIB file for the iPhone version of this game fully expanded. On the left, at item A, we see a reference to the main UIWindow of the game as well as a reference to the AppDelegate. The view controller at item B is an instance of the class RootViewController, which is a subclass of UINavigationController. The view controller Root View Controller is wired up to the Window so that it is the root controller; this causes the Root View Controller's top-level UIViewController to be displayed. In short, everything is wired up so the view at item E is displayed at launch.

In Figure 11–8, we see that there are two other UIViewControllers: Extras Controller at item C and Belt Commander Controller at item D. The object Extras Controller corresponds to the view at item G, while the object Belt Commander Controller corresponds to the view at item G. Both of these controllers have an IBOutlet in our class RootViewController so they can be referenced. In fact, many of the objects in the XIB file are connected to the RootViewController, as shown in Listing 11–1.

Listing 11–1. *RootViewController.m*

```
@interface RootViewController : UINavigationController<BeltCommanderDelegate,
UIAlertViewDelegate, GKLeaderboardViewControllerDelegate, FBSessionDelegate,
FBRequestDelegate>{
    IBOutlet UIViewController* welcomeController;
    IBOutlet BeltCommanderController* beltCommanderController;
    IBOutlet ExtrasController *extrasController;

    IBOutlet UIButton *leaderBoardButton;
    IBOutlet UIButton *tweetButton;

    UIAlertView* newGameAlertView;
    UIAlertView* pauseGameAlerTView;

    GKLocalPlayer* localPlayer;
    Facebook* facebook;

    BOOL isPlaying;
}
-(void)doPause;
-(void)endOfGameCleanup;

-(void)initGameCenter;
-(void)initFacebook;
-(void)initTwitter;

-(void)notifyGameCenter;
-(void)notifyFacebook;

-(BOOL)handleOpenURL:(NSURL *)url;

-(Facebook*)facebook;
@end
```

In Listing 11–1, we see the header for the class RootViewController. The class RootViewController is responsible for managing the application state that does not involve the actual playing of the game. This includes transitioning from view to view, handling application startup, and managing the social services from Chapter 9. As we can see in Listing 11–1, we have IBOutlets for the three UIViewControllers; welcomeController, beltCommanderController, and extrasController.

Listing 11–1 also shows that the class RootViewController has IBOutlets for two buttons: leaderBoardButton and tweetButton. A reference to these objects is required, as they will be enabled or disabled at runtime. The GKLocalPlayer is used to manage Game Center and the Facebook object is used to manage interactions with Facebook. Let's take a look at how we handle the navigation across views.

View Navigation

In order to change views when a user clicks a button, we have created a number of IBAction tasks in the implementation of RootViewController and are wired up in the XIB file to the corresponding buttons. Listing 11–2 shows two of these IBAction tasks.

Listing 11–2. *RootViewController.m (extrasButtonClicked: and playButtonClicked:)*

```
- (IBAction)extrasButtonClicked:(id)sender {
    [self pushViewController:extrasController animated:YES];
}
- (IBAction)playButtonClicked:(id)sender {
    [self pushViewController:beltCommanderController animated:YES];
    [beltCommanderController doNewGame: [extrasController gameParams]];
}
```

In Listing 11–2, we see two tasks responsible for handling a button click. The task extrasButtonClicked: is called when the user presses the Extras button on the start view. This task simply uses task pushViewController:animated:, as defined by UINavigationController, to present the view associated with extrasController. Similarly, the task playButtonClicked: uses the same task to display the view associated with beltCommandController as well as calls doNewGame: on beltCommanderController. Listing 11–3 shows how we navigate back to the start view.

Listing 11–3. *RootViewController.m (backFromExtras:)*

```
- (IBAction)backFromExtras:(id)sender {
    [self popViewControllerAnimated:YES];
}
```

In Listing 11–3, we see the task backFromExtras: that is called when the user clicks the Back button on the extras view. To navigate back to the start view, we call popViewControllerAnimated: to remove the extras view show the start view. The task popViewControllerAnimated: is defined by UINavigationController, which RootViewController is a subclass of.

There is one last piece of code that handles navigation and that is the code that response to the two dialogs that appear on the game view, as shown in Listing 11–4.

Listing 11–4. *RootViewController.m (alertView:clickedButtonIndex:)*

```
- (void)alertView:(UIAlertView *)alertView clickedButtonAtIndex:(NSInteger)buttonIndex{
    if (alertView == pauseGameAlerTView){
        if (buttonIndex == 0) {
            [self endOfGameCleanup];
            [self popViewControllerAnimated:YES];
        } else {
            [beltCommanderController setIsPaused:NO];
        }
    }
    if (alertView == newGameAlertView){
        if (buttonIndex == 0) {
            [beltCommanderController doNewGame: [extrasController gameParams]];
        } else {
            [self popViewControllerAnimated:YES];
        }
    }
}
```

In Listing 11–4, we see the task alertView:clickedButtonAtIndex:, which is defined by the protocol UIAlertViewDelegate and implemented by RootViewController. In this task, we first check which UIAlertView was clicked.

If alertView is equal to pauseGameAlertView, then we know that the user has clicked the pause button on the game view and is responding to the dialog that is presented. If the user clicks Yes on the dialog (buttonIndex == 0), then the user wants to exit their current game. We call endOfGameCleanup and popViewControllerAnimated:; the latter of course returns us the start view. The task endOfGameCleanup is responsible for reporting on the social media services and we will look at it shortly. If the user had clicked No, we simply un-pause the game calling setIsPaused: and beltCommanderController and pass in NO.

In Listing 11–4, there are two possible UIAlertViews that might be responsible for alertView:clickedButtonAtIndex: being called. We looked at the case for the UIAlertView pauseGameAlertView; the other option is that it was the newGameAlertView is responsible. This UIAlertView is displayed at the end of a game when the user has lost all of his health. This dialog allows the user to simply start a new game or return to the start view. As can be seen, starting a new game is as simple as calling doNewGame: on beltCommanderController. To return to the start view, we use the now familiar task popViewControllerAnimated:.

The code responsible for displaying the two different UIAlertView is very similar; let's just look at one of those to understand it. The code that displays the UIAlertVIew at the end of a game is shown in Listing 11–5.

Listing 11–5. *RootViewController.m (gameOver:)*

```
-(void)gameOver:(BeltCommanderController*)aBeltCommanderController{
    if (newGameAlertView == nil){
        newGameAlertView = [[UIAlertView alloc] initWithTitle:@"Your Game Is Over."
message:@"Play Again?" delegate:self cancelButtonTitle:@"Yes" otherButtonTitles:@"No",
nil];
    }
    [newGameAlertView show];

    [self endOfGameCleanup];
}
```

In Listing 11–5, we see the task gameOver: as implemented by RootViewController. The task gameOver: is defined by the protocol BeltCommanderDelegate that is the delegate protocol used to communicate between the RootViewController and the BeltCommanderController when a game starts or stops. That is to say, the RootViewController is a delegate of beltCommanderView. This relationship is defined in the XIB file. In this case, we are looking at the code that is called when the game is over. In this task we lazily create the UIAlertView newGameAlertView, specifying the text to be displayed and that self should be used as the delegate. To display the UIAlertVIew we call show, causing the view to pop-up on the user's screen. The last thing we do is call endOfGameCleanup, as shown in Listing 11–6.

Listing 11–6. *RootViewController.m (endOfGameCleanup)*

```
-(void)endOfGameCleanup{
    isPlaying = NO;
    [self notifyGameCenter];
    [self notifyFacebook];
}
```

In Listing 11–6, we see the task endOfGameCleanup that is called when the game ends, either by the user losing all her health or by quitting her active game from the pause dialog. In this task, we record that we are no longer playing by setting isPlaying to NO. The tasks notifyGameCenter and notifyFacebook inform those two services of the users score, and are described in Chapter 9.

We have looked at how the application is organized by inspecting the XIB file, understanding the flow of the application and how we implemented it. Next, we want to look at the game itself and understand how the class BeltCommanderController, along with the Actor subclasses, create the game.

Implementing the Game

We have looked at an overview of how the application is put together. We know how the views got on the screen and looked at the life cycle of the application. In this section we are going to focus on just the game mechanics. We will look at how do we get actors into the game, how they behave, and how they interact.

First we will review the classes we created throughout this book to create a simple game framework. We will then explore the class BeltCommanderController and understand how it manages the game. Lastly, we will look at each Actor and understand how its implementation makes it work the way that it does.

Game Classes

In previous chapters, we have built a collection of classes for implementing a game. We have the class GameController that is responsible for adding and removing actors from the game, as well as rendering them into the screen. We also have the class Actor that represents the characters and interactive elements of the game. Subclasses of Actor specify a Representation, a collection of Behaviors, and some custom code. The following is a review of these classes.

GameControllerThe class GameController is the beating heart of the game; subclasses of this class need only add actors and animations will start happening. To customize the GameController, subclasses should implement the task applyGameLogic to perform any game specific logic, such as add or remove actors, check for victory or failure conditions, or anything else unique to a particular game.

In addition to updating the Actors and game logic, GameController is responsible for rendering the Actors to the screen. GameController does this in two ways. First GameController extends UIControllerView so it has a primary UIView called view that can be part of an applications scene. Second, GameController has a UIView called actorsView that is the super view for all of the UIViews that represent actors in the game. By specifying the view and actorsView in a XIB file, an instance of GameController can draw to the screen. The second way that GameController is responsible for rendering the game is by working with the Representation of each Actor to create a UIView to add as a child to actorsView. For each step of the animation, GameController changes the location, scale, and transparency for each Actor's UIView. The key tasks for GameController are as follows:

- doSetup. In the task doSetup a subclass of GameController should specify the subclasses of Actor that should be kept sorted by calling setSortedActorClasses:.

- applyGameLogic. The task applyGameLogic is called once for every step of the animation. It is in this task that subclasses should implement all game logic, including

  - Setting end-of-game conditions

  - Adding and removing Actors

  - Keeping score

  - Tracking achievements

  - Updating heads-up display (HUD)

  - Managing user input (touches, gestures, and so on)

- actorsOfType:. The task actorsOfType: is used inside of the applyGameLogic to find all Actors of a particular class. The type of actor requested needs to be specified during doSetup with the setSortedActorClasses: task.

- addActor:/removeActor:. The tasks addActor: and removeActor: are how actors get added and removed from the game. These tasks collect all Actors added or removed during an animation step and then applies those changes all at once at the end of the animation loop. This prevents modification of the collections that hold the actors, so any subclass of Actor or GameController can add or remove any Actor during their apply task.

Actor

The Actor class represents anything that is dynamic in a game. In our example, the ship, asteroids, saucers, power-ups, health bars, particles, and bullets are all actors. When creating a game, a subclass of GameController will handle the big details about the game, but it is the subclasses of Actor that provide the unique and interesting behaviors of each item in the game. In practice each Actor in the game is a composition of its properties, how it is drawn (Representation), how it behaves (Behaviors), and the custom code defined in a subclass. We will look at Representations and Behaviors after we look at the class Actor. The following is a list of the properties of an Actor and what each means.

- long actorId: The actorId is a unique number assigned to each actor. This value can be used to provide a soft reference to Actors in the game.

- BOOL added/removed: The properties added and removed can be used to test if an Actor has successfully been added or removed from a game.

- CGPoint center: The property center defines the location of the center of the actor in game coordinates.

- float rotation: The rotation property indicates the rotation of the actor in radians.

- float radius: The radius of the actor describes how big it is. Together, center and radius describe the area of the game occupied by the Actor.

- NSMutableArray* behaviors: The behaviors property is a collection of Behavior objects. A Behavior object describes some sort of aspect of an actor. It might describe how it moves or when it is removed from the game. Each Behavior in the behaviors array will be applied to the actor on each step of the game.

- BOOL needsViewUpdated: Throughout a step of the game, an Actor may undergo changes that require its UIView to be updated. These might include advancing the image that should represent it or a change in state. This flag informs the GameController that work is required to synchronize the Actor with its representation.

- NSObject<Representation> representation: Each actor in the game requires an object conforming to Representation to describe how to create a UIView to represent it in the game. Representations are discussed in the next section.

- int variant: It is common to have Actors in a game that differ only slightly. Perhaps they are a different color or have a slightly different behavior. Since this is a common requirement, the property variant provides a simple way of distinguishing Actors of the same class.

- int state: The property state is used to describe the state of an Actor. This state can be anything, and is up to the subclass to define exactly what this indicates. The property state and variant work with the class ImageRepresentation to figure out which image should be used for an Actor.

- float alpha: The property alpha describes how opaque an Actor is. An Actor with an alpha of 0.0 is fully transparent, while an alpha value of 1.0 is fully opaque. Transparent regions in an Actor's representation are unaffected.

- BOOL animationPaused: The property animationPaused is used to pause any animation that is being applied to the actor. It is up to the implementation of the animation to honor this value. ImageRepresentation will not update the image, in a sequence of images, used to represent an Actor if this property is true.

In addition to the properties that define an Actor, there are a handful of tasks that come with the class Actor. These are the following:

- `-(id)initAt:(CGPoint)aPoint WithRadius:(float)aRadius AndRepresentation:(NSObject<Representation>*)aRepresentation`: This task is the designated constructor for the class Actor. It requires that each Actor has a center, a radius, and a representation. All subclasses of Actor should be sure that this task gets called when the Actor is created.

- `-(void)step:(GameController*)controller`: The task `step:` is called once per game step by the GameController. Subclasses of Actor can provide custom logic in their implementation of this task.

- `-(BOOL)overlapsWith: (Actor*) actor`: The task `overlapsWith:` is used to test if an Actor occupies any of the same space as another Actor. This can be used to implement simple collision detection.

- `-(void)addBehavior:(NSObject<Behavior>*)behavior`: The task `addBehavior:` is a utility method for adding Behaviors to an Actor in a single step.

Representation

Each Actor has to describe how the GameController should render it. The protocol Representation (defined in Actor.h) describes how a GameController can get and update a UIView for each Actor. There are two concrete implementations of Representation: ImageRepresentation and VectorRepresentation. The class ImageRepresentation uses UIImages and UIImageViews to create a UIView suitable for presenting the Actor. The class VectorRepresentation draws the Actor dynamically using Core Graphics. The details for how ImageRepresentation are covered in Chapter 6, while the implementation of VectorRepresentation is covered Chapter 7. There are two tasks defined by the protocol:

- `-(UIView*)getViewForActor:(Actor*)anActor In:(GameController*)aController`: This task is called once shortly after an Actor is added to a GameController. It is responsible for creating and returning a UIView suitable for rendering the Actor.

- `-(void)updateView:(UIView*)aView ForActor:(Actor*)anActor In:(GameController*)aController`: This task is called whenever an Actor's needsViewUpdated property is true. This task should make any changes to the UIView required to make it accurately represent the Actor.

Behavior

When creating a game, a lot of different Actors will require very similar code to implement. While it is definitely possible to construct a class hierarchy that provides just the right code for each class, I find this cumbersome. One solution to this is the Behavior protocol and the classes that implement it. Each instance of the protocol

Behavior is responsible for applying some logic to an Actor one per step. The protocol Behavior defines a single a task:

```
-(void)applyToActor:(Actor*)anActor In:(GameController*)gameController;
```

The task applyToActor:In: is implement by the classes that conform to this protocol to provide some sort of behavior to the Actors they are attached to. There are several implementations of Behavior included in the game Belt Commander, those are:

- LinearMotion: The class LinearMotion is used to move an Actor in a straight line during the game. The class provides several utility constructors to make it easy to define the motion based on a direction or a target point.

- ExpireAfterTime: The class ExpireAfterTime is used to remove an Actor after a given number of steps. This is handy for Particles, where you want to create them, add them to the scene and then forget about them.

- FollowActor: This class is used to keep one Actor at a fixed distant from anotherActor. This is used by the HealthBars attached to the Saucers to keep them together.

Now that we have reviewed the core classes use to create the game Belt Commander, it is time to look at the subclasses of the classes we just described to understand how we can extend and customize them to create the game Belt Commander.

Understanding BeltCommanderController

In previous chapters, we have extended the class GameController to make our examples. To create a whole game, we are going to do the same thing, by creating the class BeltCommanderController. The class BeltCommanderController is responsible for managing the Actors in the game, determining the end condition, updating the health and score, and interpreting input from the user. Let's take a look at the action in the game again and review how the game is played and what challenges confront the player. A screenshot of the game is shown in Figure 11–9.

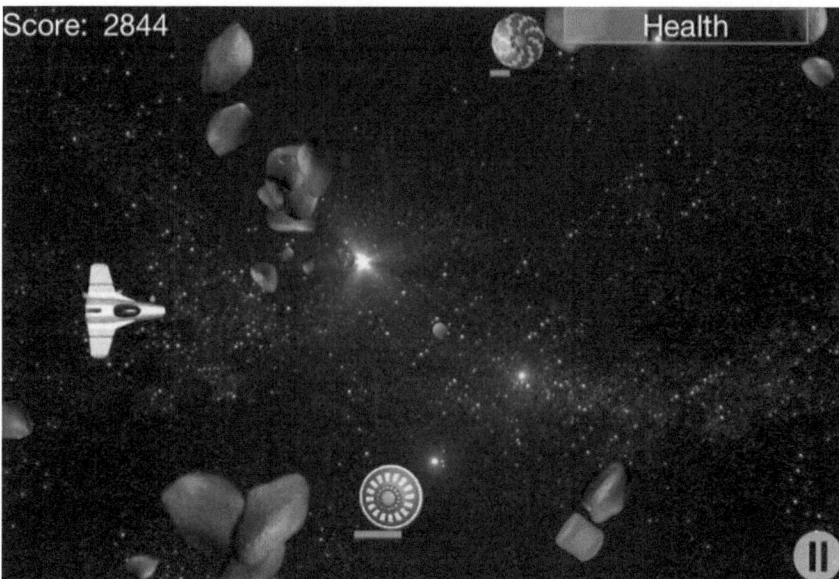

Figure 11–9. *Belt Commander, mid-game*

In Figure 11–9, we see the game Belt Commander. The player controls the ship on the left by tapping the screen. The ship will move to the same height as the tap. The ship constantly fires bullets to the right, as represented by the circle just to the lower right of the bright star in the background. From the right, moving left, come an endless wave of asteroids. If a bullet collides with an asteroid, the asteroids breaks into smaller parts, and the score is incremented. If an asteroid collides with the ship, the asteroid is destroyed, but the ship's health is decremented. To add a little chaos to the game, saucers appear on the right side of the game and travel up and down, firing bullets at the ship. The bullets will destroy asteroids, but their main purpose is to hit the ship and reduce its health. To help out our poor ship, three types of power-ups come from the left and travel to the right. If the ship is lucky enough to be in the path of a power-up, it gains its benefit. These benefits are increased health, more powerful bullets, and extra points. The game continues until the ships health is reduced to zero.

Now that we have an overview of how the game is played, we are going to look at the class BeltCommanderController, first looking at how a new game is setup and then looking at the code that runs every step of the game.

BeltCommanderController: Getting Started

To understand how this game is implemented, we need to understand how the BeltComanderController gets set up to play the game. The following section provides an overview of the initial code required to get a game up and running. Let's start with the header file for the class BeltCommanderController, as shown in Listing 11–7.

Listing 11–7. *BeltCommanderController.h*

```
@class BeltCommanderDelegate, BeltCommanderController;

@protocol BeltCommanderDelegate
@required
-(void)gameStarted:(BeltCommanderController*)aBeltCommanderController;
-(void)gameOver:(BeltCommanderController*)aBeltCommanderController;
@end

@interface BeltCommanderController : GameController{
    IBOutlet HealthBarView* healthBarView;
    IBOutlet UILabel* scoreLabel;

    GameParameters* gameParameters;
    Viper* viper;

    //achievement tracking
    int asteroids_destroyed;
}

@property (nonatomic, retain) IBOutlet NSObject<BeltCommanderDelegate>* delegate;

-(void)doNewGame:(GameParameters*)aGameParameters;
-(void)tapGesture:(UITapGestureRecognizer*)tapRecognizer;

-(void)doEndGame;
-(void)doAddNewTrouble;
-(void)doCollisionDetection;
-(void)doUpdateHUD;
-(void)checkAchievements;

-(Viper*)viper;
@end
```

In Listing 11–7, we see the header file for the class BeltCommanderController. As expected, the class extends GameController. It also defines a protocol called BeltCommanderDelegate. This protocol is implemented by RootViewController and is used to pass game state. There are two IBOutlets associated with this class. The first, HealthBarView, is the UIView subclass that is responsible for rendering the health bar at the upper right of the screen. The second, scoreLabel, is used to display the player's score as the game is played.

In Listing 11–7, we see that there is a GameParameters object called gameParameters. This object is used to determine which types of Actors should be included in the game. The idea is that this game would be free and the player could purchase different types of Actors to make it more fun (and profitable). Such in-app purchases are discussed in Chapter 10. For this chapter, we are assuming all in-app purchases have been made.

The Viper object is used to keep a reference to the ship in the game. Since we will want easy access to this object, it makes sense just to keep a pointer to it, instead of searching for it in with the other Actors. The last field is called asteroids_destroyed and used to track achievements, as described in Chapter 9.

There are a number of tasks associated with the class BeltCommanderController. The only ones we need to worry about at this point are doNewGame: and tapGesture:, the others we will cover as we go through the implementation. The task doNewGame: is called by RootViewController to start a new game—we will look at the implementation shortly. The task tapGesture: is called when a UITapGestureRecognizer detects a tap on the screen, we will see this shortly as well. More about how gestures work is covered in Chapter 8. Let's focus on the setup first.

Understanding the Setup

There are few steps that we must take to get BeltCommanderController set up. Since we know that BeltCommanderController extends GameController we know that it should implement a doSetup task, as shown in Listing 11–8.

Listing 11–8. *BeltCommanderController.m (doSetup)*

```
-(BOOL)doSetup{
    if ([super doSetup]){
        [self setGameAreaSize:CGSizeMake(480, 320)];
        [self setIsPaused:YES];

        NSMutableArray* classes = [NSMutableArray new];
        [classes addObject:[Saucer class]];
        [classes addObject:[Bullet class]];
        [classes addObject:[Asteroid class]];
        [classes addObject:[Powerup class]];
        [self setSortedActorClasses:classes];

        UITapGestureRecognizer* tapRecognizer = [[UITapGestureRecognizer alloc]
initWithTarget:self action:@selector(tapGesture:)];

        [actorsView addGestureRecognizer:tapRecognizer];
        return YES;
    }
    return NO;
}
```

In Listing 11–8, we see the task doSetup as defined by the class BeltCommanderController. In this task, we define the size of the game to be 480x320, which happens to be the native size of the iPhone screen in points. We also set the property isPaused to YES, so nothing is animating until doNewGame: is called later. The next step is to indicate which classes of Actors we want easy access to as the game progresses. In this case, we specify the classes Saucer, Bullet, Asteroid, and Powerup. Keep in mind that there is only one Viper in the game, and we have a reference to it in the header file. The last thing we do is to create a UITapGestureRecognizer and add it to actorsView. In this way, the task tapGesture: will be called when the user taps on the screen. Before any of that can happen, we need to understand how a new game is created.

A New Game

Whether we are running the first game or the hundredth, they all start with the task doNewGame:, as shown in Listing 11–9.

Listing 11–9. *BeltCommanderController.m (doNewGame:)*

```
-(void)doNewGame:(GameParameters*)aGameParameters{
    gameParameters = aGameParameters;

    [self removeAllActors];

    [self setScore:0];
    [self setStepNumber:0];
    [self setScoreChangedOnStep:0];

    viper = [Viper viper:self];
    [self addActor:viper];

    asteroids_destroyed = 0;

    [self setIsPaused:NO];
    [delegate gameStarted:self];
}
```

In Listing 11–9, we see the task doNewGame:, which is called by the RootViewController to start a new game. In this task, we set the gameParameters for future reference. Then we have to reset this object so that it is ready for a new game. To remove all Actors we call removeAllActors. We also want to reset the score, the number of steps, and the number of asteroids destroyed. We also create a Viper object and store it in the variable viper. Once we are reset, we set isPaused to No and inform the delegate that the game has started. Let's take a look at how we handle input and then we will be ready to move beyond the setup.

Handling Input

As soon as isPaused is set to NO the game is up and running. At this point, the only Actor in the game is the viper. Let's take a moment to understand how we intercept input to make it move before the asteroids and aliens show up. The task tapGesture: is shown in Listing 11–10.

Listing 11–10. *BeltCommanderController.m (tapGesture:)*

```
-(void)tapGesture:(UITapGestureRecognizer*)tapRecognizer{
    if (![self isPaused]){

        CGSize gameSize = [self gameAreaSize];
        CGSize viewSize = [actorsView frame].size;
        float xRatio = gameSize.width/viewSize.width;
        float yRatio = gameSize.height/viewSize.height;

        CGPoint locationInView = [tapRecognizer locationInView:actorsView];
```

```
        CGPoint pointInGame = CGPointMake(locationInView.x*xRatio,
locationInView.y*yRatio);
        [viper setMoveToPoint: pointInGame within:self];
    }
}
```

In Listing 11–10, we see the task tapGesture: that is called when the user taps the screen. In this task, after checking if we are paused, we have to convert the point of the touch into game coordinates. On the iPhone, this is not strictly necessary, because the view actorsView is the same number of points wide and high as the size of our game. However, when we play this game on the iPad, the actorsView is 1024x682, while our game is still 640x480. In order to convert from one coordinate space to the other, we divide the game width by the width of actorsView and multiply it by the X location of the touch. We repeat the process for the height and Y value. Once we have the point figured out we call setMoveToPoint:within: on viper, which sets it in motion. We will look at the implementation of setMoveToPoint:within: when we take a closer look at the class Viper in the next section.

We have covered all there is to know about getting BeltCommanderController setup to play the game. Everything is ready to start processing user input and adding new Actors to the mix. The following section describes how BeltCommanderController checks for an end condition, adds Actors, and other task appropriate for the BeltCommanderController class.

BeltCommanderController One Step At a Time

We have our BeltCommanderController class all set up and ready to play the game. We are going to look at the code that is run for every step to manage the overall state of the game. We will look at how the end condition is tested for, how we add new Actors to the scene, and how we manage the interactions between the Actors. Since BeltCommanderController is a subclass of GameController, we start out in game logic in the task applyGameLogic, as shown in Listing 11–11.

Listing 11–11. *BeltCommanderController.m (applyGameLogic)*

```
-(void)applyGameLogic{
    if ([viper health] <= 0.0f){
        [self doEndGame];
    } else {

        [self doAddNewTrouble];
        [self doCollisionDetection];
        [self doUpdateHUD];

        if ([self stepNumber]%30 == 0){
            [self checkAchievements];
        }
    }
}
```

In Listing 11–11, we see the task applyGameLogic that is called once for every step of the game. The first thing we do is check to see if the viper's health is below 0.0. If it is, we end the game—simple as that. If the viper is still alive, we continue by calling doAddNewTrouble, doCollisionDetection, and doUpdateHUD. The call to checkAchievements is done only once every 30 steps. This is done for performance reasons, because we don't really want to make these extra calls for every step. This does have the drawback that achievements might be achieved by the player but not recorded. For a more sophisticated game we would break up which achievements are tested when.

We just listed off a good number of tasks, and we will take each in turn, starting with the task doEndGame, as shown in Listing 11–12.

Listing 11–12. *BeltCommanderController.m (doEndGame)*

```
-(void)doEndGame{

    [self setIsPaused:YES];
    [delegate gameOver:self];
}
```

In Listing 11–12, we see the very simple task doEndGame where we simply pause the game and call gameOver: on the delegate. The task gameOver: is implemented by RootViewController and can seen in Listing 11–5. Next, let's review how Actors are added to the game.

Adding Actors

We know from experience that Actors are added by calling addActor:. However, to make a game we have to include some logic that dictates when and how an Actor is added. Let's continue and see how we add new Actors to the game in the task doAddNewTrouble, as shown in Listing 11–13.

Listing 11–13. *BeltCommanderController.m (doAddNewTrouble)*

```
-(void)doAddNewTrouble{
    if ([gameParameters includeAsteroids] && arc4random() % (5*60) == 0){
        if ([[self actorsOfType:[Asteroid class]] count] < 20){
            [self addActor:[Asteroid asteroid:self]];
        }
    }
    if ([gameParameters includeSaucers] && arc4random() % (10*60) == 0){
        if ([[self actorsOfType:[Saucer class]] count] < 3){
            [self addActor:[Saucer saucer:self]];
        }
    }
    if ([gameParameters includePowerups] && arc4random() % (20*60) == 0){
        [self addActor:[Powerup powerup:self]];
    }
}
```

In Listing 11–13, we see the task doAddNewTrouble that is responsible for adding new Actors to the game. For each of three types of Actors we might add, we first check to see if gameParameters indicates that we should add it. If gameParameters indicates that we should add an Actor of a particular type, we perform a random check to see if we actually add it. For example, Asteroids have a 1 in 300 chance to be added. To add each Actor we create it and call addActor:. The Actor's constructor handles the details of how and where each Actor is created. We will now move on and examine how Actors interact with each other.

Collision Detection

Now we know how each Actor is added to the scene. Let's move on and look at how they interact. If we consider the five main Actors in this game, there are several interactions that must be considered, as follows:

- Bullets and Asteroids

- Asteroids and the Viper

- Bullets and the Viper

- Bullets and Saucers

- Power-ups and the Viper

These interactions are handled in the task doCollisionDetection, the first part of which is shown in Listing 11–14.

Listing 11–14. *BeltCommanderController.m (doCollisionDetection, part 1)*

```
-(void)doCollisionDetection{
    NSSet* bullets = [self actorsOfType:[Bullet class]];
    NSSet* asteroids = [self actorsOfType:[Asteroid class]];
    NSSet* saucers = [self actorsOfType:[Saucer class]];
    NSSet* powerups = [self actorsOfType:[Powerup class]];

    for (Asteroid* asteroid in asteroids){
        for (Bullet* bullet in bullets){
            if ([bullet overlapsWith:asteroid]){
                [bullet decrementDamage: self];

                asteroids_destroyed++;
                [asteroid doHit:self];
                [self incrementScore: [asteroid level]*10];
                break;
            }
        }
        if ([asteroid overlapsWith:viper]){
            [viper decrementHealth: [asteroid level]*2];

            Shield* shield = [Shield shieldProtecting:viper From: asteroid];
            [self addActor:shield];

            asteroids_destroyed++;
            [asteroid doHit:self];
        }
    }
```

In Listing 11–14, we see the first part of the task doCollisionDetection where we handle the interaction concerning Asteroids. The first thing we do in this task is to get a reference to all of the different Actors we are going to need to interact with. Once we have reference to all of the Asteroids, Bullets and the Viper we can start testing to see if there are collisions. Inside the outermost loop in Listing 11–14, we start by testing to see if each Asteroid overlaps with each Bullet. If they do, we have the bullet decrement its damage, which usually removes it from the game, but more on that later. We also record that we have destroyed another Asteroid before calling doHit; on the Asteroid. Lastly, we increment the score.

In Listing 11–14, after we have considered the relation between each Asteroid and Bullet, we check to see if the Asteroid is overlapping the Viper. If it is, we decrement the health of the Viper and add a Shield to the scene. The Shield is an Actor that is just decoration and has no game function beyond that. It makes it look like a shield went up to protect the spaceship.

Let's look at the rest of the doCollisionDetection task and understand how the rest of the inter-Actor relationships are handled. The task continues in Listing 11–15.

Listing 11–15. *BeltCommanderController.m (doCollisionDetection, part 2)*

```
for (Bullet* bullet in bullets){
    if ([[bullet source] isKindOfClass:[Saucer class]]){
        if ([viper overlapsWith: bullet]){
            [viper decrementHealth: [bullet damage]];

            Shield* shield = [Shield shieldProtecting:viper From: bullet];
            [self addActor:shield];

            [self removeActor:bullet];
            break;
        }
    } else {
        for (Saucer* saucer in saucers){
            if ([saucer overlapsWith: bullet]){
                [saucer decrementHealth:[bullet damage]];
                Shield* shield = [Shield shieldProtecting:saucer From:bullet];
                [self addActor:shield];
                [self removeActor:bullet];
                break;
            }
        }
    }
}

for (Powerup* powerup in powerups){
    if ([powerup overlapsWith:viper]){
        [powerup doHitOn:viper in:self];
    }
}
}
```

In Listing 11–15, we continue looking at the task doCollisionDetection. While iterating over all of the Bullets, we start by considering the interactions between Bullets and the Vipers. We do this by checking the source property on bullet. If the source was a Saucer, then the Bullet was intended for the Viper and cannot hurt a Saucer. If the Bullet overlaps with the Viper we decrement the Viper's health, add a Shield, and remove the Bullet from the game.

If the Bullet was not shot by a Saucer, it must have come from the Viper. So we check to see if bullet overlaps with any of the Saucers. If it does, we decrement the health of the Saucer, add a Shield to the Saucer, and remove the Bullet.

The last relationship we have to consider in doCollisionDetection is that between Powerups and the Viper. We iterate over all of the Powerups in the game and check if they overlap. If they do we call doHitOn:in: on the Powerup to apply the bonus. The implementation of doHit:in: will be described when we consider the Powerup class in detail.

We have now looked at all of the interactions between the different actors, before look at each of the different Actor classes, let's take a look how we update the score and the health bar on the screen.

Updating the HUD

For every step of the game, the score or the health meter might change. Collectively, these two components are called the heads-up display (HUD). Looking back to Listing 11–7, we see that these two components are connected to the BeltCommanderController though an IBOutlet. We see that the scoreLabel is a simple UILabel, but the class HealthBarView is unfamiliar. We will discuss the class HealthBarView in a moment; first, let's look at the code that updates these components in the task doUpdateHUD, as shown in Listing 11–16.

Listing 11–16. *BeltCommanderController.m (doUpdateHUD)*

```
-(void)doUpdateHUD{
    if ([self stepNumber] == [self scoreChangedOnStep]){
        [scoreLabel setText: [[NSNumber numberWithLong:[self score]] stringValue] ];
    }
    [healthBarView setHealth:[viper health]/[viper maxHealth]];
}
```

In Listing 11–16, we see the task doUpdateHUD that is called for every step of the game. In this task, we check to see if the current stepNumber is equal to the property scoreChangedOnStep. The property scoreChangedOnStep is the value of stepNumber when the score was last updated. So, if stepNumber is equal to scoreChangedOnStep, we know that the score was changed in this step and hence, scoreLabel needs to be updated. To update scoreLabel, we simply call setText: and pass in an NSString version of the score.

In Listing 11–16, we also call setHealth: on healthBarView to update how the health bar is drawn. The object healthBarView is of type HealthBarView, which is simply a UIView with a custom drawRect: task, as shown in Listing 11–17.

Listing 11–17. *HealthBarView.m (drawRect:)*

```
- (void)drawRect:(CGRect)rect
{
    [self setDefaults];
    int index = 0;

    float marker = [[percents objectAtIndex:index] floatValue];
```

```
    while (percent > marker) {
        marker = [[percents objectAtIndex:++index] floatValue];
    }

    UIColor* baseColor = [colors objectAtIndex:index];
    const float* rgb = CGColorGetComponents( baseColor.CGColor );


    UIColor* frameColor = [UIColor colorWithRed:rgb[0] green:rgb[1] blue:rgb[2]
alpha:.8f];
    UIColor* healthColor = [UIColor colorWithRed:rgb[0] green:rgb[1] blue:rgb[2]
alpha:.5f];

    [frameColor setStroke];
    [healthColor setFill];

    CGSize size = [self frame].size;
    CGContextRef context = UIGraphicsGetCurrentContext();

    CGRect frameRect = CGRectMake(1, 1, size.width-2, size.height-2
                                 );
    CGContextStrokeRect(context, frameRect);

    CGRect heatlhRect = CGRectMake(1, 1, (size.width-2)*percent, size.height-2);
    CGContextFillRect(context, heatlhRect);
}
```

In Listing 11–17, we see the task drawRect: of class HealthBarView. In this task, we draw two rectangles. The first rectangle is filled and the second is just an outline. Besides just drawing the rectangles, we want to adjust the color of the rectangles to reflect how damaged the ship is. To figure out which color we should use as our baseColor, we iterate through the NSArray percents until the percent we are going to draw is higher than percent at index.

Once we have our base color, we create two modified colors from it. The first is the color frameColor used for the outline of the health bar; it has an alpha of 0.8. The second color is healthColor and will be used to draw the interior of the health bar; it has an alpha of 0.5.

To draw the rectangles we first define them by calling CGRectMake. Once the CGRect is created we call CGContextStrokeRect and CGContextFillRect respectively.

We have looked at the class BeltCommanderController and know it works. In the next section, we will consider the five main Actors and understand how they have the features that they do.

Implementing the Actors

We have reviewed how the class BeltCommanderController handles the big picture. It controls how and when Actors are created and manages the interactions between them. However, a lot of the functionality of the game is implemented in the classes that make up the different Actors. In this section, we will take each of the four main Actors and understand what the key customizations are to make them work the way we want. In several earlier chapters, we have implemented Actors of different types. In that light, we are only going to look at the bits of code that make each interesting feature of the Actor work. If you would like a refresher on how to implement Actors, check out Chapters 6 and 7. Also, you can always look at the full source code. Let's start with the class Viper.

The Viper Actor

The Viper class represents the spaceship at the left side of the screen. This is the Actor that the user controls to achieve a high score. The Viper fires Bullets continuously to the right. To move the Viper, the user taps the screen indicating a point above or below the Viper where it should travel. When traveling the graphic changes to show thrusters on the top or bottom of the craft. Also, a Viper can intercept a Powerup that improves its guns. Lastly, the Viper has a fixed amount of health that slowly regenerates. In summary, the key features of the Viper are as follows:

- Moves to a point specified by user.

- Thrusters fire when ship moves.

- Fires bullets continuously.

- Improves quality of bullets when "powered up."

- Viper has a health value that regenerates.

Let's take a look at the source code and see how we implement these features, starting with the header file for context, as shown in Listing 11–18.

Listing 11–18. *Viper.h*

```
enum{
    VPR_STATE_STOPPED = 0,
    VPR_STATE_UP,
    VPR_STATE_DOWN,
    VPR_STATE_COUNT
};

@interface Viper : Actor <ImageRepresentationDelegate, LinearMotionDelegate>{
    LinearMotion* motion;

    long lastStepDamageWasModified;
}
@property (nonatomic) float health;
@property (nonatomic) float maxHealth;
@property (nonatomic) long lastShot;
@property (nonatomic) long stepsPerShot;
@property (nonatomic) BOOL shootTop;
@property (nonatomic) int damage;

+(id)viper:(GameController*)gameController;
-(void)setMoveToPoint:(CGPoint)aPoint within:(GameController*)gameController;
-(void)incrementHealth:(float)amount;
-(void)decrementHealth:(float)amount;
-(void)incrementDamage:(GameController*)gameController;
```

```
-(void)decrementDamage:(GameController*)gameController;
@end
```

In Listing 11–18, we see the header file for the class Viper. In this file, we see that we define an enum with three states. This is very much like the enums we have declared in the past to specify the state of an Actor. Looking at the properties, we can see that we have a property for health and maxHealth. We also have a property called lastShot, which will be used to keep track of whether it is time to shoot another Bullet in conjunction with the property stepsPerShot. The BOOL property shootTop is used to keep track if the next Bullet should come from the top of the Viper or not. The last property damage indicates the damage value of the next Bullet.

The tasks listed in 11–18 are pretty mundane. There are four tasks for incrementing and decrementing health and damage. Each keeps the value of health and damage within a reasonable range. The interesting task is setMoveToPoint:within:, as shown in Listing 11–19.

Listing 11–19. *Viper.m (setMoveToPoint:within:)*

```
-(void)setMoveToPoint:(CGPoint)aPoint within:(GameController*)gameController{
    if (motion){
        [[self behaviors] removeObject: motion];
    }
    CGPoint point = CGPointMake([self center].x, aPoint.y);

    if (point.y < self.center.y){
        [self setState:VPR_STATE_UP];
    } else if (point.y > self.center.y){
        [self setState:VPR_STATE_DOWN];
    }


    motion = [LinearMotion linearMotionFromPoint:[self center] toPoint:point
AtSpeed:1.2f];
    [motion setDelegate:self];
    [motion setStopAtPoint:YES];
    [motion setPointToStopAt:point];

    [motion setStayInRect:YES];
    CGSize gameSize = [gameController gameAreaSize];
    [motion setRectToStayIn:CGRectMake(63, 31, 2, gameSize.height - 63)];

    [motion setWrap:NO];

    [[self behaviors] addObject:motion];

}
```

In Listing 11–19, we see the task setMoveToPoint:within:. In this task, the goal is to set up a LinearMotion object that will move the Viper to the specified point. The first step is to see if we have an existing LinearMotion object stored at the variable motion. If so, we want to remove it from our active set of behaviors. The next step is to create a new CGPoint called point based on the CGPoint that was passed in. Because this method will accept a point from anyplace in the game, we have to find a point that keeps the Viper horizontally fixed. Once we know where out point is, we can set our state to either VPR_STATE_UP or VPR_STATE_DOWN.

Creating the new LinearMotion is as simple as passing the Viper's current center and point to the constructor linearMotionFromPoint:toPoint:AtSpeed:. Setting self as the delegate and telling it to stop when it reaches the specified point further modifies the LinearMotion. The last modification is to set LinearMotion so that it stays within the specified rectangle. This prevents the Viper from going outside the bounds of the screen.

We have looked at how the Viper moves about the screen. The rest of the interesting features are implemented in the task step: as shown in Listing, 11–20.

Listing 11–20. *Viper.m (step:)*

```
-(void)step:(GameController*)gameController{
    long stepNumber = [gameController stepNumber];
    if (stepNumber - lastShot > stepsPerShot - damage){

        CGPoint center = [self center];
        CGSize gameSize = [gameController gameAreaSize];

        float offset;
        if (shootTop){
            offset = -15;
        } else {
            offset = 15;
        }

        CGPoint bCenter = CGPointMake(center.x + 20, center.y + offset);
        CGPoint bToward = CGPointMake(gameSize.width, bCenter.y);

        Bullet* bullet = [Bullet bulletAt:bCenter TowardPoint:bToward From:self];
        [bullet setDamage:damage];
        [gameController addActor:bullet];

        shootTop = !shootTop;
        lastShot = stepNumber;
    }
    if (stepNumber % 60 == 0){
        [self incrementHealth:1];
    }
    if (lastStepDamageWasModified + 10*60 == stepNumber){
        [self decrementDamage:gameController];
    }
}
```

In Listing 11–20, we see the task step:, which is called for every step of the game. We do a number of things in this task; primarily, we figure out if a Bullet should be shot and, if so, shoot it. But we also regenerate our health and keep track of whether we are still under the influence of a Powerup. To figure out if we should fire a Bullet, we see if stepNumber minus lastShot is greater than stepsPerShot minus damage. In this way, we fire more often if damage is higher, which happens when a Powerup has given the Viper extra damage.

To fire a Bullet, we figure out if the Bullet should come from the top or bottom of the ship. Then it is just a matter of creating the CGPoints that describe its motion. Lastly, we clean up by flipping the value of shootTop and set lastShot to stepNumber.

In Listing 11–20, we also increase our health. This done by calling incrementHealth every 60 frames. Calculating if we should decrement our damage is done by checking when our damage was last increased and calculating if 600 frames (~10 seconds) have passed.

All in all, very little is required to get our Viper class to do what we want. Let's move on the Asteroid class.

The Asteroid Class

The Asteroids are the primary opponents in Belt Commander (presumably, the word "belt" in the title refers to an asteroid belt). These Actors start on the right and travel to the left. When they reach the left side, they start over again on the right. We have used the variations of the Asteroid class several times in this book, and they are well documented. However, the feature that makes the Asteroid interesting is how it breaks up into multiple smaller asteroids while shooting Particles. This code was modified from previous examples, and is shown in Listing 11–21.

Listing 11–21. *Asteroid.m (doHit:)*

```
-(void)doHit:(GameController*)controller{
    if (level > 1){
        int count = 1;
        float percent = arc4random()%1000/1000.0f;
        if (percent > 0.9){
            count = 3;
        } else if (percent > 0.5){
            count = 2;
        }
        for (int i=0;i<count;i++){

            float radius = [self radius];

            float rx = arc4random()%1000/1000.0*radius*2 - radius;
            float ry = arc4random()%1000/1000.0*radius*2 - radius;
            CGPoint newCenter = CGPointMake(self.center.x + rx, self.center.y + ry);

            Asteroid* newAst = [Asteroid asteroidOfLevel:level-1 At: newCenter];
            [controller addActor:newAst];
        }
    }

    int particles = arc4random()%4+1;
    for (int i=0;i<particles;i++){
        ImageRepresentation* rep = [ImageRepresentation
imageRepWithDelegate:[AsteroidRepresentationDelegate instance]];
        Particle* particle = [Particle particleAt:self.center WithRep:rep Steps:25];
        [particle setRadius:6];
        [particle setVariant:arc4random()%AST_VARIATION_COUNT];
        [particle setRotation: (arc4random()%100)/100.0*M_PI*2];
```

```
            LinearMotion* motion = [LinearMotion linearMotionRandomDirectionAndSpeed];
            [particle addBehavior:motion];

            [controller addActor: particle];
        }
        [controller removeActor:self];
}
```

In Listing 11–21, we see the task doHit: as implemented by the class Asteroid. In this task, we check to see if the Asteroid we are working with is above level one. If it is, then we need to create sub-Asteroids to replace this one. In other examples we just randomly picked a number of child Asteroids to create. However, while play-testing Belt Commander, this was unsatisfactory, because the number of Asteroids generated was way too high. To reduce the number of Asteroids created, a percentage was given for each possible value of count. So only 10 percent of the time will an Asteroid create three children Asteroids. 40 percent of the time there will be two children Asteroids. The rest of the time only a single Asteroid is created. In this way the game was balanced and kept fun and playable.

The other interesting thing about the task doHit:, as shown in Listing 11–21, is how it creates Particles to emphasize the destruction of the parent Asteroid. By creating 1–5 Particles that look just like Asteroids and sending them out in random directions, we create a very pleasing Asteroid destruction.

Let's continue and explore a little about the class Saucer.

The Saucer Class

The Saucer class is mostly a troublemaker. It appears at either the top or bottom of the screen, shooting Bullets at the Viper. What makes the Saucer class troublesome is that the Bullets most often break apart Asteroids, creating an unpredictable pattern. The implementation of the Saucer class in this chapter is much likes its implementation in the other chapters. However, some modification was made to the step: task as shown in Listing 11–22.

Listing 11–22. *Saucer.m (step:)*

```
-(void)step:(GameController*)gameController{
    if (health <= 0){
        [gameController incrementScore:[self radius]*10];
        [gameController removeActor:self];
        return;
    }

    CGSize gameAreaSize = [gameController gameAreaSize];
    CGPoint center = [self center];

    if (center.y < -[self radius]){
        [linearMotion setDirection:DIRECTION_DOWN];
    } else if (center.y > gameAreaSize.height + [self radius]){
        [linearMotion setDirection:DIRECTION_UP];
    }
```

```
    if (arc4random()%180 == 0){
        BeltCommanderController* bc = (BeltCommanderController*)gameController;

        [gameController addActor:[Bullet bulletAt:[self center] TowardPoint:[bc
viper].center From:self]];
    }
}
```

In Listing 11–22, we see the implementation of the class step: from the class Saucer. In this task, we check to see if the Saucer's health is below zero. If it is, we increment the gameControllers score and remove the Saucer from the game. The next part of the task step: is to determine if the Saucer should change direction on account of going off the screen. The last section of the task step: is responsible for shooting Bullets. In order to do this, we get the viper property of the gameController so we can target the Bullet on it. This is the first case where we had to cast gameController to BeltCommanderController. It raises the question of how we might add a general way to reference particular actors (like the viper) without depending on class definition. But, this is an exercise for another time.

The last Actor we are looking at is Powerup.

The Powerup Class

The Powerup class is pretty simple, really. It is just an Actor that moves from the right to the left. Have reviewed in other chapters how it appears to be spinning and how it flashes at the end of its life. In this game, the interesting thing is the effect it has on a Viper. This feature is implemented in the task doHitOn:in:, as shown in Listing 11–23.

Listing 11–23. *Powerup.m (doHitOn:in:)*

```
-(void)doHitOn:(Viper*)viper in:(GameController*)gameController{
    if (self.variant == VARIATION_HEALTH){
        [viper incrementHealth:30];
    } else if (self.variant == VARIATION_DAMAGE){
        [viper incrementDamage:gameController];
    } else if (self.variant == VARIATION_CASH){
        [gameController incrementScore:1000];
    }
    [gameController removeActor:self];
}
```

In Listing 11–23, we see the task doHitOn:in: of the class Powerup. This is called when a Powerup collides with the Viper. The implementation could not be more straightforward: we simply check the variant of the Powerup and apply the correct effect. For VARIATION_HEALTH, we increment the Viper's health by 30. For VARIATION _DAMAGE, we increment the Viper's damage. Lastly, for VARIATION_CASH, we just increment the score by 1,000. I think the simplicity of this task is a direct byproduct of using Actors the way we do. It turned out to be a strong OO model.

Summary

In this chapter, we brought all of the techniques we have discussed in this book together. We started by exploring what our game, Belt Commander, is and how it is played. We then reviewed the flow of the application and looked at how we used the `UINavigationController` class to implement switching between views. We continued by looking at the class `BeltCommanderController` and learned how to implement the major parts of the game. Lastly, we looked at a few specific implementation choices that added life to our `Actors`.

When creating a game, there will always be unique code and techniques involved. Each game is one of a kind and requires unique effort to bring to life. It would be impossible to write a book that explains every possible challenge a game developer can face. However, many developers and game designers are going to face requirements that are very similar to what I have outlined here. It is my hope that this book shows how to work through a number of these common challenges. From the most basic issue of switching screens and saving user data to the more complex topics like particle animations and in-app purchases, I think of this book as a map of the code and classes that are the infrastructure a game. Ultimately it is up to each developer to fill in the middle parts and create a new game. Thank you for reading this book.

Designing and Creating Graphics

In this appendix, we'll cover three main topics: designing graphics, creating graphics, and some open source tools to use in doing so.

When creating a video game, a lot of the work is put into creating the graphics used in the game. In this appendix, we will be considering 2D games. These graphics will be images for the actors in the game, backgrounds, buttons, and other decorations. These images are a large part of the game; they are the first thing people see, and they can dictate whether people are going to download the game. Images are what grab people's attention. It is the image of Link or Mario that people remember—not the collision detection algorithm. It is not to say that there are not popular games with simple graphics, because there definitely are. But even creators of successful games with simple graphics spend a lot of time to make sure their art conforms to a unified design goal.

In addition to concerns of style, a game designer must also consider what is required to create images suitable for building a game. We must consider the size of the final images as well as how the images are organized. Should we use separate images for everything, or can we create composite image for backgrounds? How do we organize our images so we can easily make changes? Do we have a budget of how many megabytes of images we can include? These are just a couple of the questions I will be addressing in this Appendix.

The last part of this appendix will be a quick review of some of the great open source tools available to a game designer. These tools are not only important for small game developers, but they were critical in the writing of this book.

The Art in Video Games

Before a video game is created, someone has an idea about how the game is played and what it looks like. That initial vision may or may not make it into the final release. In any case, it is ideal to have a single person, or group of people, responsible for shaping the art of the game so it has a consistent style. Once a game has art direction, work will commence on building the assets for the game. As the game comes together, other pieces of artwork will be created to support the game, such as a web site or an advertisement. These other pieces of work should be in the same visual style. In this section, we will do a quick review of what art style is and how that shapes users expectations.

Style in Video Games

The style of a video game refers to all of the choices the designers and artists made to create the art in a game, and how that art fits together. The style of a game is no different than style in other products. For example, consider a bottle of teriyaki sauce. It would not be unreasonable for the word *teriyaki* to be drawn in a typeface that looks like it came from the Far East, perhaps like the word in Figure A–1.

TERIYAKI

Figure A–1. *"Teriyaki" rendered in a Far East style*

Figure A–1 shows a word rendered with a style that matches the expectations of the users of teriyaki sauce. The many makers of teriyaki sauce set this expectation over the years. Not all products have a socially expected style to draw from. In fact, if you are making a new game in a new genre, users may have no expectation at all about the style. If you are making a game about knights and castles, the user will welcome a style that looks medieval. In any case, it is important that the style of the art be consistent throughout the game.

Having a consistent style does not require the art to be overly elaborate. The goal is simply to make the art appear to be created by a single hand, as the example in Figure A–2 shows.

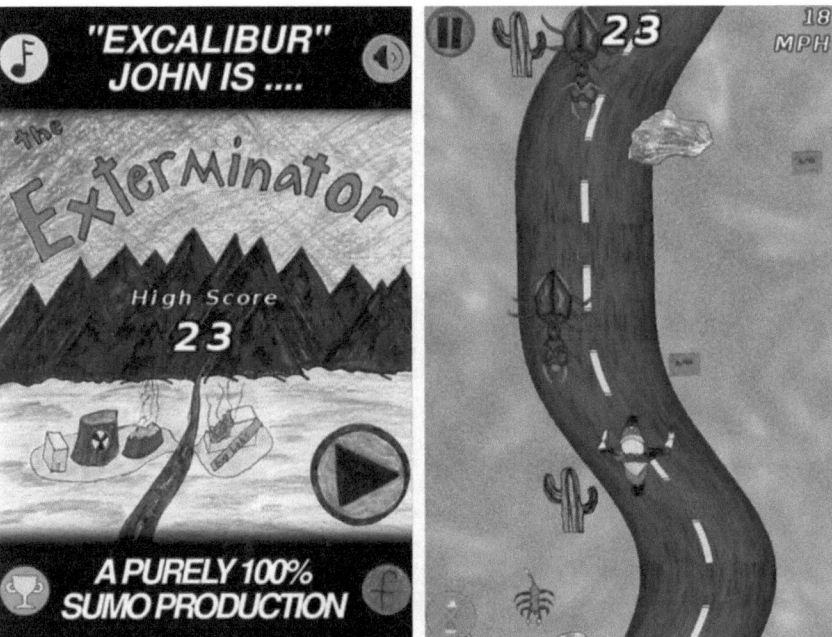

Figure A-2. *The Exterminator, by SUMO Productions. © 2011 Table Mountain Church*

On the left side of the figure, we see the Welcome screen for the game. Phil Hassey developed this simple game with a number of teens from Table Mountain Church (who own the copyright). Because this game was collaboration between Phil and a number of young people without years of graphics or game design experience, they adopted a simple hand-drawn style. This choice allowed several people with limited drawing skills to contribute art to the game. In order to maintain the style throughout the game, we can see that four buttons, in the corners of the left screen, are also drawn by hand. The bottom right button is a Facebook button. By making their own Facebook button, they maintain a consistent style. If they used one of the stock Facebook buttons, it would look out of place from the entire visual experience. We should also consider a game that does not have consistent style, as shown in Figure A-3.

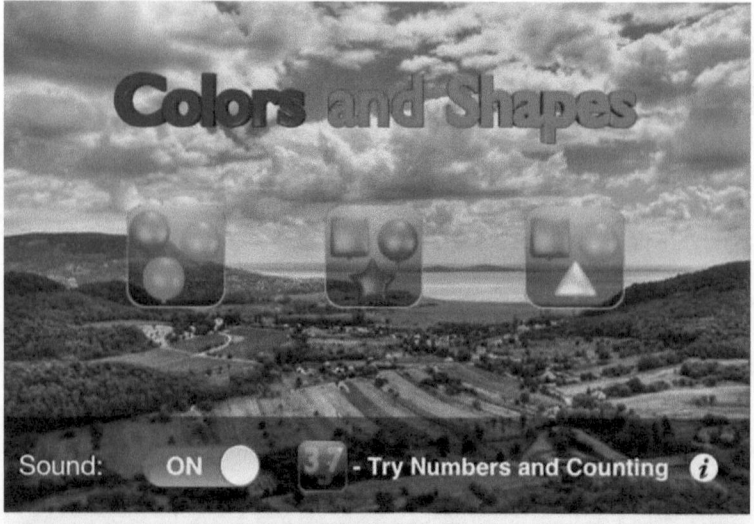

Figure A-3. *Colors and Shapes. © ClayWare Games. Background by txd on Flickr*

In Figure A–3, we see an old game I created for the original iPhone called Colors and Shapes. On the left is the Welcome screen, where a child or parent can pick one of three activities. For each activity, the player is directed by voice and text to perform a number of simple tasks involving colored balloons. The 3D text and the balloons were rendered with Blender 3D, and I think are pretty consistent. They both use strong primary colors and have a sense of depth—not a perfect match, but pretty close. However, I don't think the background was a good choice. While it is a very nice photo, it doesn't match the balloons at all, and is visually distracting.

TIP: Creative Commons has a license option where all you have to do to use a creative work in your game is include a byline in your credits. Flickr allows you to search for images that are published with this license. We found the background image for Colors and Shapes this way.

Branding and Perception

As mentioned, the art for a game is more than just images moving about the screen. People remember the art and associate it with the game. Figure A–4 shows an advertisement created for the game Space Attack!!.

Figure A–4. *Space Attack ad*

This flyer was handed out at a convention. When I made this flyer, I was in the middle of updating the art for the game, so I decided to use the new art for the flyer, even though the game had not been updated in the App Store with the new art. So when people downloaded the game, they downloaded the version with the old art, as shown in Figure A–5.

Figure A–5. *Space Attack!!, old art*

Because this art did not match the art on the flyer, and the art on the game is considerably worse, people felt ripped off, even though the game was free. I failed to appreciate my users' expectations, and paid for it with some negative reviews. This could have easily been avoided by either getting the game updated in time for the conference or using the older graphics in the ad.

We have taken a quick look at what style in a video game is and why it is important. Let's now shift our focus to the more practical matters of creating the image files. After that, we will review a number of excellent tools for creating 2D art.

Creating the Images Files

Regardless of the art direction of your game, at the end of the day you will need a number of images files, at a specific size, to actually implement your game. We are going to first explore some naming conventions that will help you manage the case where you want different images for different devices. Then we will review the images required by all iOS applications—things like icons and launch screens. We will conclude this section with a discussion of how to pick appropriate image sizes.

Naming Conventions

At the time of writing, there are three different screen resolutions that must be accounted for when creating an application. It is very common to want to use different images based on which one of the three resolutions the application is running on. To support these resolutions, a naming convention was created to facilitate this in a mostly painless way. The idea is that by naming the images in a particular way, a different image will be loaded depending on the device. So there is no need to test which device you are using in your code. The pattern for this naming convention is

```
Basename[@2x][~iphone|~pad].extension
```

So basename can be whatever you want—for example, "background." You can then optionally append an @2x to indicate that image is for a high-resolution displays, like the retina display of the iPhone 4. After indicating whether the image is a high-resolution version, you can specify a device class, either iPhone or iPad. Lastly, you add the extension as you normally would. Table A–1 shows how this naming convention works for an image that would otherwise be called background.png.

Table A–1. *Naming Convention For a Background Image in Portrait*

Device	Filename	Resolution of image in Pixels
iPhone 3GS, iPod Touch 3· Generation	background~iphone.png	320x480
iPhone 4, iPod Touch 4· Generation	background@2x~iphone.png	640x960
iPad 1, iPad 2	background~ipad.png	768x1024
iPad 3*	background@2x~ipad.png	1536x2048

* As of this writing, there is no iPad 3. I am making a logical conjecture about the naming convention, assuming the iPad 3 has a retina display like the iPhone, starting with version 4.

In Table A–1, we see in the first column a set of devices for each row. In order to provide a full screen, portrait background image for each set of devices, we name our background file as shown in the second column. The right-most column shows the resolution of the image in pixels. In the sample code for this appendix, the background image of the application is configured to support the devices available in the emulator. I have changed each background image to include text indicating which file is being loaded. Figure A–6 shows the different devices with the different images.

Figure A–6. *A different background for each device*

On the left of the figure, we see the iPhone running with the same display that is found on the iPhone 3GS. In the middle, we see a screen size as if it was running on the iPhone 4, relative to the 3GS. On the right is the screen size for the first and second iPad. Listing A–1 sets the background image for this application.

Listing A–1. *ViewController.m (viewDidLoad)*

```
- (void)viewDidLoad
{
    [super viewDidLoad];

    UIImage* image = [UIImage imageNamed:@"background"];
    [backgroundImageView setImage:image];
}
```

In Listing A–1, we see the `viewDidLoad` task of the class `ViewController`. We load a `UIImage` by calling `imageNamed` and passing in just the base name. What is returned is the image appropriate for the specific devices based on the filenames of the images included in the project. In this case, we have the same filenames as those shown in Table A–1.

Now let's take a look at how we specify other support images based on device type, such as icons and splash screens.

Support Images

Every iOS application has a number of images that must accompany the application. At the very least, an icon must be specified. Optionally, an image can be specified that is displayed as the application is launched. It is generally a good idea to provide a launch image, especially if your application takes a moment to start. Specifying which image to use is very simple: Xcode provides places where images can be dragged to specify icons and launch images, as shown in Figures A–7 and A–8.

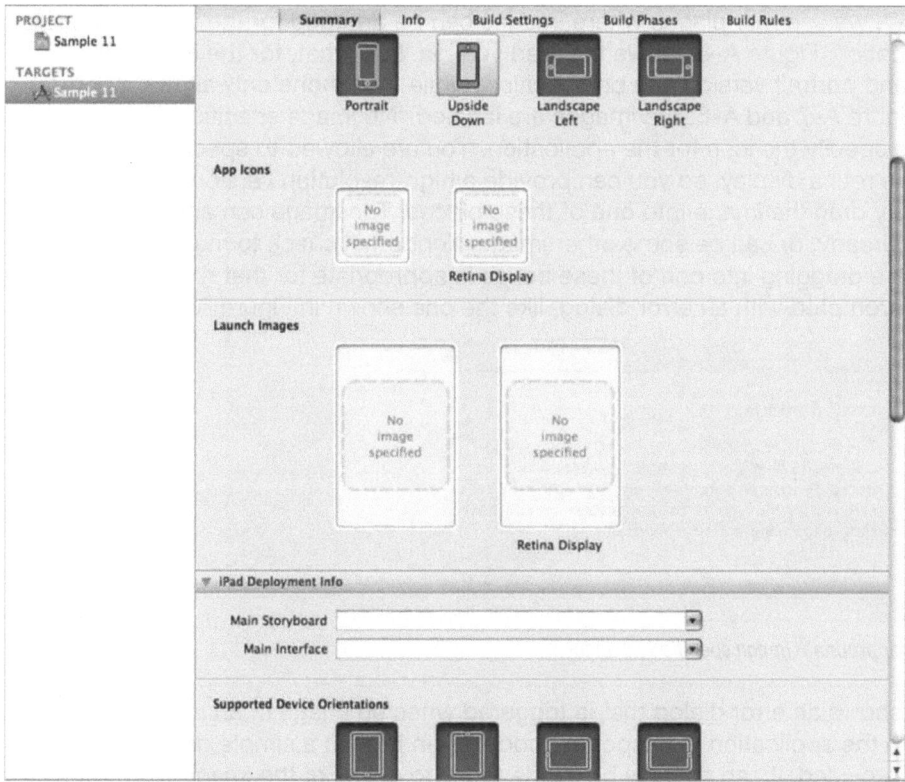

Figure A–7. *Specifying icons and launch images for the iPhone*

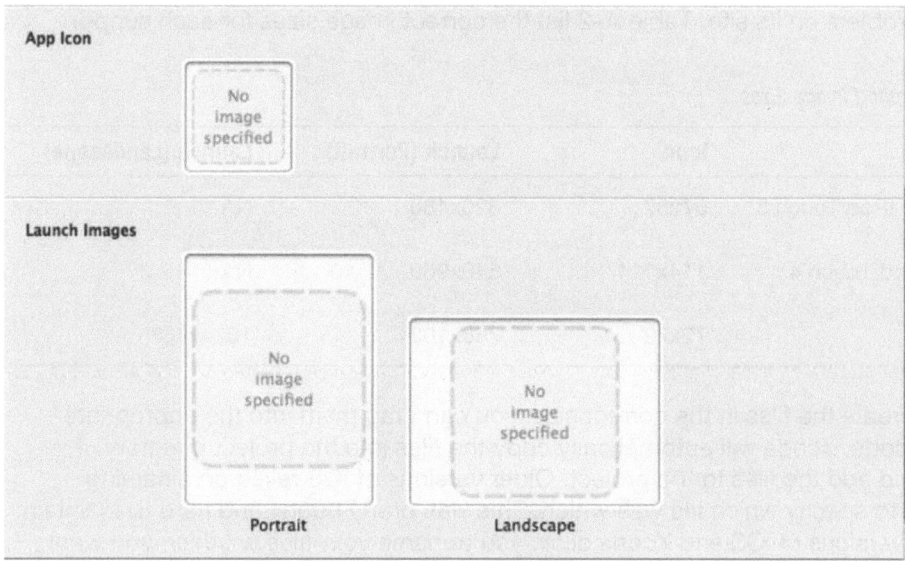

Figure A–8. *Specifying icons and launch images or the iPad*

In Figure A–7, we see four places where images can be dragged for the iPhone version of the application; Figure A–8 shows the iPad version. Note that, for the iPad, a landscape and portrait version can be specified, while the iPhone only allows a portrait image. In Figure A–7 and A–8, the images are labeled "No image specified." The top two allow you to specify the icon for the application. You are allowed to specify a different image for the retina display, so you can provide a high-resolution version. To specify an image, simply drag the image into one of these places. The image can either be part of the project already, or can be some other image. Xcode will check to make sure the image you are dragging into one of these boxes is appropriate for that role. If it is not, you will be prompted with an error dialog, like the one shown in Figure A–9.

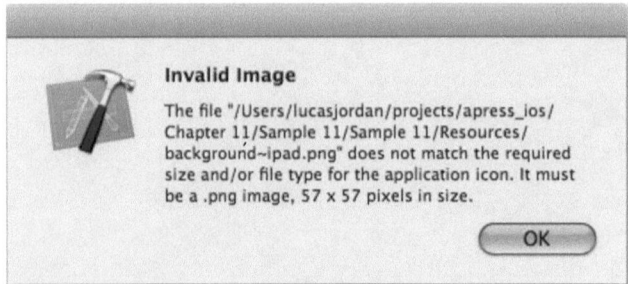

Invalid Image

The file "/Users/lucasjordan/projects/apress_ios/ Chapter 11/Sample 11/Sample 11/Resources/ background~ipad.png" does not match the required size and/or file type for the application icon. It must be a .png image, 57 x 57 pixels in size.

OK

Figure A–9. *Inappropriate image specified as an icon*

Figure A–9 shows an error dialog that is triggered when an image of an incorrect size is dragged into the application icon spot in Xcode. Even though a simple drag-and-drop interface is provided, it can actually be a little tricky to get all of the images formatted correctly. This is tricky because there is incorrect information about image sizes on most web sites, including Apple's. I hope that by the time you are reading this, Apple has fixed this problem on its site. Table A–2 list the correct image sizes for each support image.

Table A–2. *Support Image Sizes*

Device	Icon	Launch (Portrait)	Launch (Landscape)
iPhone 3GS, iPod Touch 3	57x57	320x480	NA
iPhone 4, iPod Touch 4	114x114	640x960	NA
iPad 1 + 2	72x72	768x1024	1024x768

Once you create the files in the correct size, you can drag them into the appropriate spots on Xcode. Xcode will automatically copy the files into the project directory (if required) and add the files to the project. Older versions of iOS relied on a naming convention to specify which file was which. This was pretty buggy and hard to maintain. The newer versions of iOS and Xcode allow you to name your files anything you want. Listing A–2 shows how this information is represented in our `info.plist` file.

Listing A–2. *Sample A–Info.plist*

```
<key>CFBundlePrimaryIcon</key>
        <dict>
                <key>CFBundleIconFiles</key>
                <array>
                        <string>my_icon_file.png</string>
                        <string>my_icon_file_retina.png</string>
                        <string>my_icon_file_ipad.png</string>
                </array>
                <key>UIPrerenderedIcon</key>
                <false/>
        </dict>
```

In Listing A–2, we see that the filenames are simply stored in an array. The names of files are the names I gave them, and can be anything—they are not names assigned by Xcode.

We have looked at two examples of how we can have a single logical image, but have a few different versions of it at different resolutions. The following section talks briefly about how best to create the same image at different resolutions.

Mutli-Resolution Images

There are several cases where an application has two different images files, at different resolutions, that are being used as the same logical image in the application. This is done so that the application can provide the best user experience based on which device the user has. The ideal situation for any application is displaying images at a 1:1 ratio (number pixels in the image to the number of pixels on the screen). We want to avoid scaling where possible, because the scaling produces lower quality images and is an extra and potentially expensive operation.

Apple computers seem to really excel at scaling images. If you have the opportunity, try playing an old-fashioned DVD (not a high-definition video) on one of the new 27–inch iMacs. When played in full-screen mode, the images are scaled up a lot, yet still look excellent. Whatever this magic scaling algorithm is, it seems to be included iOS devices as well. As a result, it is tempting to simply use high-definition images for everything and let iOS scale the images down when the application is running on a lower resolution device. This will work; the quality of the scaled images is very good. However, when it comes to animations, it really starts to pay off to provide those smaller images for the low-res device. Remember that the iPhone 3GS not only has a smaller screen, it is less of a computer than the iPhone 4. By having the iPhone 3GS scale down the images, you are asking a device with less computing power to do more. It is worth the trouble to maintain two sets of images in most cases.

When creating a high-resolution and low-resolution set of images, we could just create the high-resolution set and create a script that scales down each one to make a low-res version. This strategy will definitely prevent the slower device from having to do the scaling, but it can hurt quality, especially with text. If your art assets are part of a rendering process of any kind, say from a 3D representation or from vector art, it makes sense to render each image twice, once for high-resolution and then again for low.

A Multi-Resolution Example

Let's look at an example. Say you are creating an icon for your application with Adobe Illustrator (or similar application). If you create a new file that is 57 pt x 57 pt, you can export this file at 72 dpi and get an image that is 57 pixels x 57 pixels, perfect for the low-res iPhone icon. You can then export this image again at 144 dpi and get an image that is 114 pixels x 114 pixels, which is perfect for the retina version of the iPhone. To make the icon for the iPad, you can export again at 92 dpi, making an image that is 72 pixels x 72 pixels. Figure A–10 shows the difference between rendering at different scales verses scaling down.

Figure A–10. *Rendering verses scaling*

In Figure A–10, we see three different versions of an icon. In the middle is a vector presentation, nice and smooth. On the left is the image produced by Illustrator when asked to export a 57x57 version of the image in the middle. The image looks pixelated because it is scaled up so we can see the quality of the rendering. The image on the right is the result of Illustrator creating an 114x114 image and then having GIMP scale it down to 57x57 (Sinc Lanczos3). Even this simple two-step process introduces noise. Notice the lighter pixels around the dot of the i on the right. Maybe in this case, it doesn't matter too much, but it seems to me if you are going to spend the time (and money) to create high-quality art, it is worth going the extra mile and preventing this type of issue.

Keeping with the ideas presented here about quality and rendering, the following section walks you through an example of how to select the final image size for an application.

Creating Final Assets

We have taken a look at how to represent images with multiple resolutions and why that is important. But when it comes to game art, we still have not answered the question, how big should my images be? In this section, we will take an example game that runs on both the iPad and both versions of the iPhone and figure out the best sizes for images used in that game. We will be using the game from Chapter 3, Coin Sorter.

Because we are working with three possible image size—one for the iPad and two for iPhone—we should look at the relative sized (in pixels) of these screens and get sense of the differences in size. Figure A–11 shows the different screen size.

Figure A–11. *Relative resolution of the different screens*

In Figure A–11, we see a rectangle representing the screen size for each of the iOS devices on the market. The left-most image is the size of the iPhone 3GS (and its two predecessors). The second rectangle from the left shows the size of the iPhone 4, which has four times as many pixels as the iPhone 3GS. The third rectangle represents the size, in pixels, of the iPad 1 and iPad 2. Note that the iPhone 4 has almost as many pixels as the iPad 1 and iPad 2. The rectangle all the way on the right is most probably the resolution for the next iPad 3 (or whatever it will be called). Let's take a look at how these different sizes affect our decisions when it comes to laying out our game.

Screen Real Estate

Though we have three different screen sizes in terms of pixels, we only have two different aspect ratios. The aspect ratio of the two iPhones is the same at 2:3, while the aspect ratio of the iPad is 3:4. Working with these differences, we want to figure out how to lay out our applications to make the best use of the screen space. Let's start by looking at the layout of the game from Chapter 3, as shown in Figure A–12.

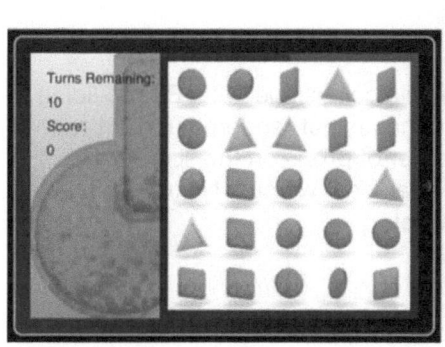
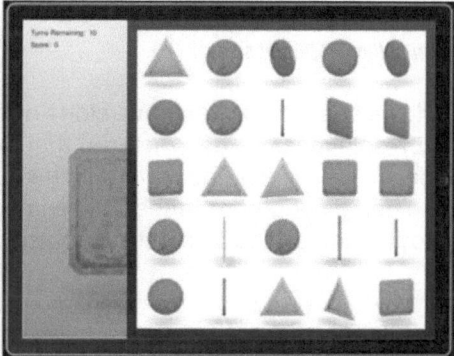

Figure A–12. *Coin Sorter on iPhone (retina) and iPad*

In Figure A–12, we see on the left the game running on the iPhone in landscape, and on the right we the game running on the iPad in landscape. In both cases, we use the game area on the right surrounded by a grey border. Because the game area is square, we only have to figure out one dimension, but we still have to figure out the best sizes for our coin images based on how much space the game area takes up. Figure A–13 shows the size of these two areas and how it relates to size of coins.

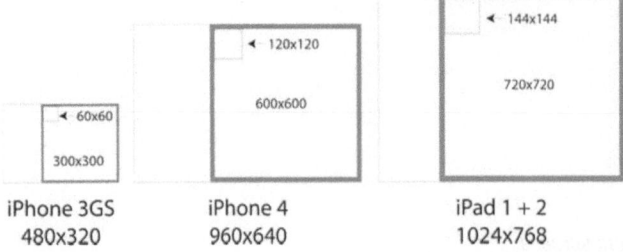

iPhone 3GS
480x320

iPhone 4
960x640

iPad 1 + 2
1024x768

Figure A–13. *Relative size of game area and coin images in pixels*

In Figure A–13, we see the screen size of our three devices. For the two iPhone versions, the white game area with the thick grey border are the same size in terms of points, 300x300. Of course, the iPhone 4 version 600x600 pixels versus the 300x300 pixels of the iPhone 3GS, but because we are concerned with finding the best size for our images, we are concerned with pixels and not points. The size of the game area on the right for the iPad 1 and iPad 2 is 720 pixels (and points). So, to calculate the size of each coin image, we simply divide by 5. We get 60x60 for the 3GS, 120x120 for the iPhone 4, and 144x144 for the iPads. Figure A–14 shows one of the coin images in all three sizes.

Figure A–14. *Coin in three different sizes*

In Figure A–14, we see three different versions of the file coin_circle0029. By providing these different resolutions, we have insured that our images get rendered at a 1:1 ratio for the number of pixels in the image versus the number of pixels on the screen. This assures we are doing our part to render these images as best we can. Next, let's take a look at some of the tools used to create graphical assets.

Tools

The following is short description of three open source graphics tools that can be used to create compelling graphical assets. These are the tools I have used to create the majority of the art files for this book. Not only are these tools open source, they run on Windows, OS X, Linux, and probably other platforms as well.

GIMP

The GNU Image Manipulation Program (or GIMP) is an open source image editing application. It provides Adobe Photoshop-like functionality with the low, low price of free. During game development, it is necessary to have an image manipulation program for many tasks. Figure A–15 shows an example of the GIMP.

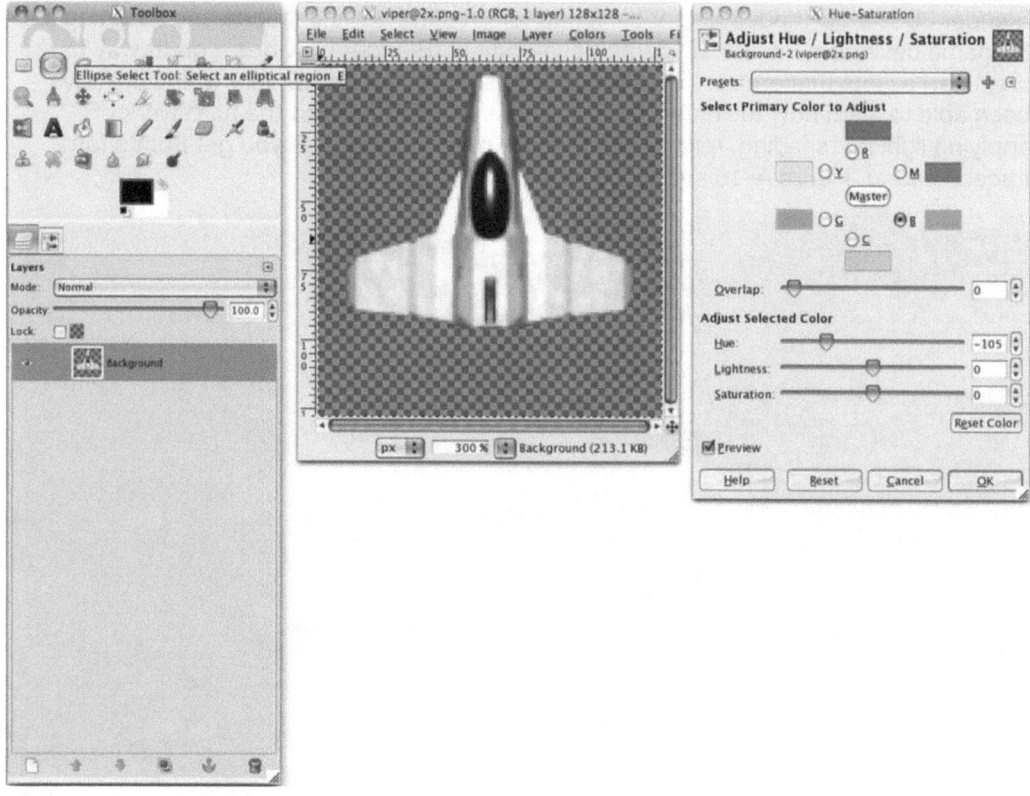

Figure A–15. *Adjusting the color of a spaceship in GIMP*

In Figure A–15, we see our spaceship from previous chapters open in GIMP. GIMP allows you to adjust the color of an image to make variations of it. In Figure A–15, we started with a ship that had blue highlights and have converted them to green, perhaps for a multiplayer scenario.

WIth GIMP (www.gimp.org) you can

- View images
- Crop images
- Resize image
- Adjustment color

- Combine images
- Fine-tune images
- Perform the other thousand image related tasks in game development

Blender 3D

Blender 3D is a 3D modeling tool designed for creating 3D animations and assisting in 3D game development. For 2D game development, Blender provides a great way to create art assets that are very consistent. I personally can't draw very well, but I have been able to learn how to "draw" objects in Blender and let it do the hard work of applying lighting, shading, reflections, and the other cool effects you get from a ray traced imaged. Figure A–16 shows Blenders interface.

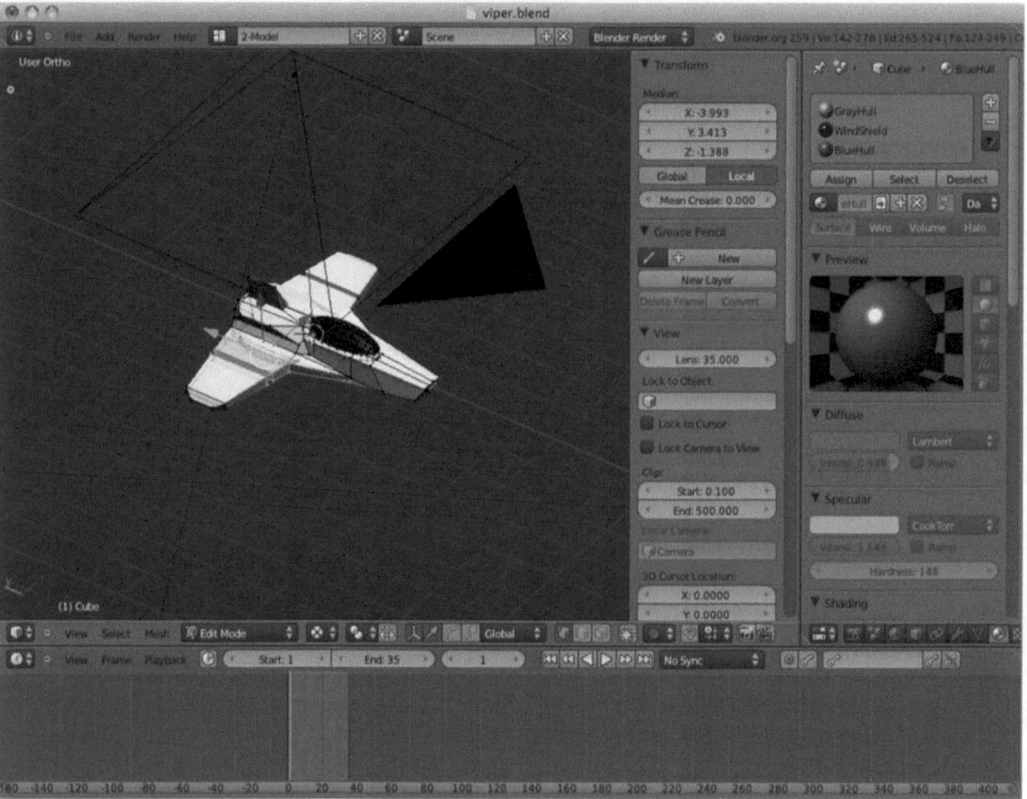

Figure A–16. *Blender 3D*

In Figure A–16, we see the UI for Blender. I will be honest: there is a lot to learn about Blender. But it is very rewarding to master each new technique with this application. The folks who make Blender recently revamped the entire application, making it easier to use. If you tried Blender in the past and got frustrated with it, give it another shot. It is a lot easier to use.

With Blender (www.blender.org) you can

1. Create 2D images to be used with as actors in game

2. Pre-render high-quality movies for cut scenes

3. Render high-quality background images

4. Render sequences of image to created animated actors

5. Create isometric tile sets

Inkscape

Inkscape is a vector drawing application. Although most games won't do a lot of vector drawing at runtime, vector-editing applications are very good at doing WYSIWYG layout. For laying out all of the great art you created with GIMP and Blender, Inkscape can't be beat. Figure A–17 shows the Inkscape application.

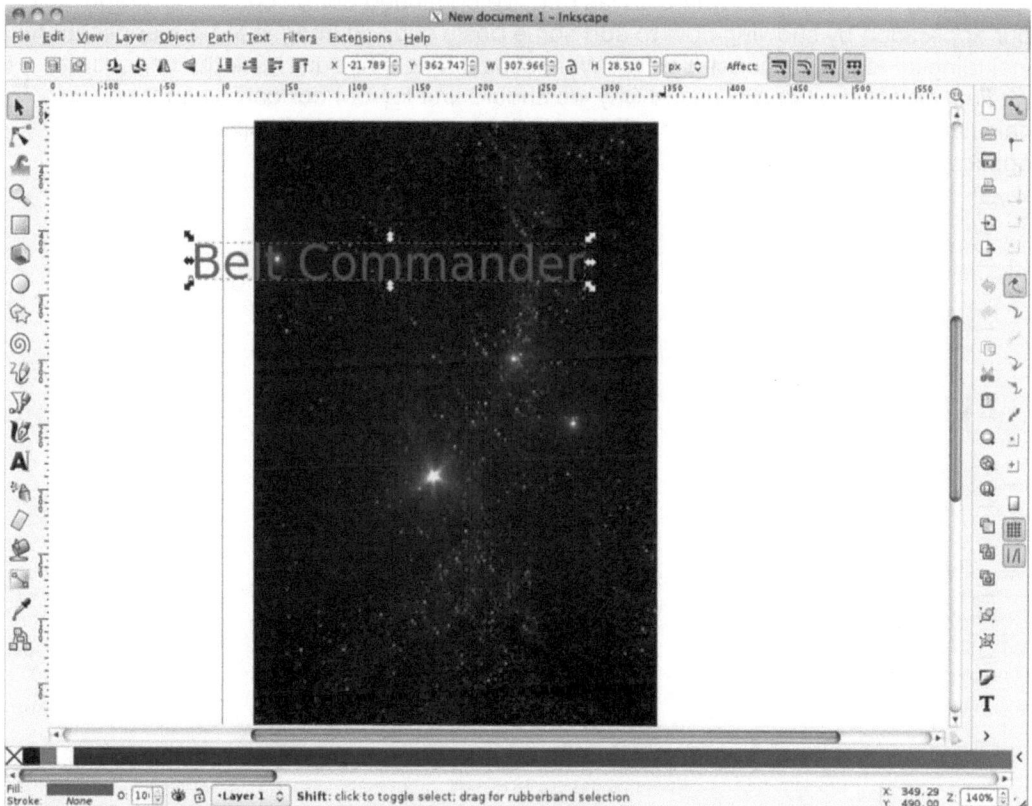

Figure A–17. *Inkscape*

The figure shows my early lay out work on the Welcome screen for the application Belt Commander. As you can see, I have imported the star field image used throughout this book and am starting to add the title text.

With Inkscape (`http://inkscape.org`) you can

1. Create text images

2. Lay out images and text relative to each other

3. Create Welcome screens, Helps screens, and other static content

4. Build wireframes and prototypes

Summary

We have reviewed some simple concepts regarding style in our games. This appendix highlighted the importance of consistent style by showing examples where style was maintained and examples where it was not. We continued our discussion of graphics in our application by showing how to add some basic support images to your application, including an icon and launch screen. We discussed how to determine the correct size for final art assets. We discussed why we should first lay out our game and then calculate the size that each item will be. Lastly, we reviewed three open source applications that are very useful for creating content for your game.

Index

H

I, J